Willis's Elements of Quantity Surveying

Willis's Elements of Quantity Surveying

Twelfth Edition

Sandra Lee
MSc, FRICS, MCIOB

William Trench
FRICS

Andrew Willis
BSc, FRICS, DipArb, FCIArb

WILEY Blackwell

This edition first published 2014

Registered Office
John Wiley & Sons, Ltd, The Atrium, Southern Gate, Chichester, West Sussex, PO19 8SQ, United Kingdom.

Editorial Offices
9600 Garsington Road, Oxford, OX4 2DQ, United Kingdom.
The Atrium, Southern Gate, Chichester, West Sussex, PO19 8SQ, United Kingdom.

For details of our global editorial offices, for customer services and for information about how to apply for permission to reuse the copyright material in this book please see our website at www.wiley.com/wiley-blackwell.

Library of Congress Cataloging-in-Publication Data

Lee, Sandra Jean, 1953–
Willis's elements of quantity surveying / Sandra Jean Lee, William Trench,
James Andrew Willis. – 12th Edition.
pages cm
Based on an original work: Elements of quantity surveying / Arthur James Willis. 1935.
Includes bibliographical references and index.
ISBN 978-1-118-49920-7 (pbk.)
1. Building–Estimates. 2. Quantity surveying. I. Willis, Andrew, 1957– II. Trench, William, 1937–
III. Title. IV. Title: Elements of quantity surveying.
TH435.W685 2014
692′.5–dc23

2013030752

A catalogue record for this book is available from the British Library.

Wiley also publishes its books in a variety of electronic formats. Some content that appears in print may not be available in electronic books.

Cover image: courtesy of Shutterstock.com
Cover design by Sandra Heath

Set in 10.5/13pt Minion by SPi Publisher Services, Pondicherry, India
Printed and bound in Malaysia by Vivar Printing Sdn Bhd

1 2014

Contents

Preface

This book was first published in 1935, and in the preface to that first edition it stated that it was intended 'to be a book giving everything in its simplest form and to assist a student to a good grounding in first principles'. Each successive edition has been brought up to date; however, we have always striven to maintain the original guiding principles, which are as relevant today as they were 70 years ago.

Whilst the use of the traditional bill of quantities continues to decline and today is only one of a variety of options open to the industry for the procurement of construction contracts, nevertheless, the skills of measurement are still very much required in some form or another under most procurement routes.

This edition recognises the publication by the Royal Institution of Chartered Surveyors (RICS) of the second volume of the *New Rules of Measurement – Detailed Measurement for Building Works* (NRM2), and the text has been updated accordingly.

The basic structure of the book generally follows that of previous editions, setting down the measurement process from first principles and assuming the reader is coming fresh to the subject.

Whilst it is recognised that modern computerised measurement techniques utilising standard descriptions might appear far removed from traditional taking-off, it is only by fully grasping such basic principles of measurement that they can be adapted and applied to alternative systems. It is for this reason that the examples continue to be written in traditional form.

The book opens with an overview of the need for measurement and the differing rules governing measurement at different stages of the design or project cycle. The main focus of the book remains on the detailed measurement of elements of a building using the rules from NRM2 and concludes with guidance on how to use the data collected during the measurement process to create the tender documents.

Whilst the role of the quantity surveyor is subject to continual change, we hope that students will find this book as useful as their predecessors have.

Sandra Lee
William Trench
Andrew Willis

Acknowledgements

We are indebted to Ruth Pearson, who prepared the drawings. Her efforts are gratefully acknowledged.

Abbreviations

a.b.	as before
a.b.d.	as before described
agg.	aggregate
BCIS	Building Cost Information Service
BS	British Standard
CAWS	Common Arrangement of Work Sections
c/c	centres
ddt	deduct
diam.	diameter
d.p.c.	damp proof course
d.p.m.	damp proof membrane
EDI	Electronic Data Interchange
e.w.s.	earth work support
ex	out of
hw.	hardwood
JCT	Joint Contracts Tribunal
n.t.s.	not to scale
PC	Prime Cost
MC	Measurement Code
n.e.	not exceeding
NRM	*New Rules of Measurement – Order of cost estimating and elemental cost planning*
r.c.	reinforced concrete
RIBA	Royal Institute of British Architects
RICS	Royal Institution of Chartered Surveyors
r.w.p.	rainwater pipe
SMM	Standard Method of Measurement
sw.	softwood
swg	standard wire gauge

Chapter 1
Introduction

The modern quantity surveyor

The training and knowledge of the quantity surveyor have enabled the role of the profession to evolve over time into new areas, and the services provided by the modern quantity surveyor now cover all aspects of procurement, contractual and project cost management. This holds true whether the quantity surveyor works as a consultant or is employed by a contractor or subcontractor. Whilst the importance of this expanded role cannot be emphasised enough, success in carrying it out stems from the traditional ability of the quantity surveyor to measure and value. It is on the aspect of measurement that this book concentrates.

The need for measurement

There is a need for measurement of a proposed construction project at various stages from the feasibility stage through to the final account. This could be in order to establish a budget price, give a pre-tender estimate, provide a contract tender sum or evaluate the amount to be paid to a contractor. There are many construction or project management activities that require some form of measurement so that appropriate rates can be applied to the quantities and a price or cost established.

The general approach adopted in this book is to concentrate on the traditional approach to construction whereby the client will employ a designer, and once the design is complete the work is tendered through the use of bills of quantities. Other procurement approaches move the

Willis's Elements of Quantity Surveying, Twelfth Edition. Sandra Lee, William Trench and Andrew Willis.

Published 2014 by John Wiley & Sons, Ltd.

need for detailed measurement to later stages of the project cycle and away from activity undertaken by the client's team to that of the contractor's team.

The need for rules

The need for rules to be followed when undertaking any measurement becomes clear when costs for past projects are analysed and elemental rates or unit rates are calculated and then applied to the quantities for a proposed project. For greater accuracy in pricing, it is important to be able to rely consistently on what is included in an element or unit, and this helps build a more reliable cost database.

Following the Royal Institute of British Architects (RIBA) 2013 Work Stages, the measurement undertaken at Stage 1 – 'Preparation' –needs to be of basic areas or functional units, and the guidelines of the Royal Institute of Chartered Surveyors (RICS) *Code of Measuring Practice* are commonly followed. This enables comparisons to be made between different schemes and options when assessing the feasibility of a project.

When preparing a cost plan, the need to include the same items in each element is important so that costs for that element can be accurately applied. In May 2009, the RICS published the first in its planned new set of rules for measurement dealing with the order of cost estimates and elemental cost planning. The RIBA work stages and the *New Rules of Measurement* (NRM) are explained further in Chapter 2.

The same need for rules applies when measuring for bills of quantities. If a document is to be used for tender purposes and included in a contract, then the contractor needs to know the basis of the measurement and what is included or excluded from an item to be priced. Historically, standard methods of measurement have been used to provide these rules and are available in various forms worldwide. The RICS NRM – detailed measurement for building works (NRM2) – have now been published and are part of the RICS 'black book' guidance for accepted practice in the United Kingdom. At post-contract stages, it is important that the rules used in the contract document (if applicable) are followed to minimise disputes.

Establishing the approach

The approach to take for any measurement is to decide its purpose and the level of design detail available, enabling the adoption of the most appropriate rules and procedures.

Chapter 3 will look at the early stages of a building project, and the remainder of the book will then focus on the detailed measurement for bills of quantities. Having an ability to read and understand the rules for measurement for bills of quantities should enable the measurer to appreciate the requirements of different rules and approaches.

Chapter 2
Detailed Measurement

Method of analysing cost

It is evident that if a building is divided up into its constituent parts, and the cost of each part can be estimated, an estimate can be compiled for the whole work. It was found in practice that by making a 'schedule' setting out the quantity of each item of work for a project, the labour and material requirements for these could be more readily assessed. This schedule at RIBA stage 5 can be in the form of a bill of quantities which, when priced by a contractor, provides a tender sum for a project. It must not be forgotten that a traditional bill of quantities only produces an estimate. It is prepared and priced before the erection of the building and gives the contractor's estimated cost. Such an estimated cost, however, under the most commonly used construction contracts, becomes a tender and a definite price for which the contractor agrees to carry out the work as set out in the bill. The bill must, therefore, completely represent the proposed work so that a serious discrepancy between actual and estimated cost does not arise.

Origin of the bill of quantities

Competitive tendering is one of the basic principles of most classes of business, and if competitors are given comprehensive details of the requirements it should be fair to all concerned. However, historically when tendering based on drawings and specification, builders found that considerable work was involved in making detailed calculations and measurements to form the basis for a tender. They realised that by getting

Willis's Elements of Quantity Surveying, Twelfth Edition. Sandra Lee, William Trench and Andrew Willis.

Published 2014 by John Wiley & Sons, Ltd.

together and employing one person to make these calculations and measurements for them all, a considerable saving would be made in their overhead charges. They began to arrange for this to be done, each including the surveyor's fee for preparing the bill of quantities in their tender, and the successful competitor paying. Each competing builder was provided with the same bill of quantities which could then be priced in a comparatively short space of time. It was not long before this situation was realised by the architect and employer. Here the employer was paying indirectly for the quantity surveyor through the builder, whereas the surveyor could be used as a consultant if a direct appointment was made. This would give the employer greater control over the amount paid to the surveyor and the opportunity to increase the service that was provided. In this way, the quantity surveyor began to get the authority of the employer and was employed to prepare a bill of quantities for tendering purposes.

The measurement process

The main purpose of a bill of quantities is therefore for tendering. Each contractor tendering for a project is able to price the work on exactly the same information with a minimum of effort. This gives rise to the fairest type of competition.

Despite the demise of bills of quantities on large projects, over 50% of the value of all building work in the United Kingdom is still let using lump-sum contracts with firm or approximate quantities. Most other procurement routes, such as design and build and management contracting in its various forms, also involve quantification of the work in some form or other by the main contractor, subcontractor or package contractor, and therefore the measurement process continues to be of importance.

Computerised and other alternative measurement systems have become more and more widely used. However, it is only by having a detailed understanding of the traditional method of setting down dimensions and framing descriptions that such systems can be fully understood and properly utilised. There continues to be development of 3D computer-aided design software that integrates with building information modelling, and the potential exists to generate quantities directly from the computer model. These software programs, however, have difficulty in producing quantities in accordance with any standard method and as yet have not removed the need for the quantity surveyor to prepare the tender and contract documents.

Attributes of a quantity surveyor

What, then, are the desired measurement skills of a good quantity surveyor? An ability to describe clearly, fully and precisely the requirements of the designers and arrange the bill of quantities so that the contractor's estimator can quickly, easily and accurately arrive at the estimated cost of the work is essential. This being so, it is obviously important that the surveyor should be able to write clearly in language that will not be misunderstood, and have a sound knowledge of building materials and construction and of customs prevailing in the industry. Moreover, the surveyor must be careful and accurate in making calculations, have a systematic and orderly mind and be able to visualise the drawings and details.

Divisions of bill preparation

The traditional preparation of a bill of quantities divides itself into two distinct stages:

(1) The measurement of the dimensions and the compilation of the descriptions from the drawings and specification. This process is commonly known as *taking-off*.

(2) The preparation of the bill. This involves the calculation of volumes, areas etc. (*squaring the dimensions*). Traditionally, this was followed by entering the descriptions and the squared dimensions on an abstract (*abstracting*) to collect similar items together and present them in a recognised bill order. From this abstract, the draft bill was written (*billing*).

Through the utilisation of computerised systems, the various stages have become more integrated. The facility now exists for direct input of quantities and formulation of descriptions through the use of standard libraries of descriptions, and the lengthy collating and bill preparation processes are carried out automatically. It should be noted that there is often still the need to produce preliminaries and preambles separately and to input uncommon items (*rogue* items) that are peculiar to the particular project. Checking total quantities and careful editing of the bill are still required to identify any data entry errors.

Quantities as part of the contract

Where a contract with quantities is used for a project, the bill of quantities forms one of the contract documents, with the contract providing that the quantity of work comprised in the contract shall be that set out in the bill of quantities. In such a case, the contractor is expected to carry out and the employer to pay for neither more nor less than the quantities given, an arrangement that is fair to both parties. Thus, it will be seen how important accuracy is in the preparation of the bill, and how a substantial error might lead an employer to enter into a contract that involved a sum considerably beyond that contemplated.

So that contractors appreciate how the quantities have been prepared and what is included in each item, the quantity surveyor uses the rules from a standard method of measurement. In the United Kingdom, the current standard method used for measuring building works is *NRM2*. A bill of quantities will be interpreted by a number of contractors in competition, and it must therefore be complete and a suitable basis for a contract.

Contractor-produced quantities and estimates

The subject of quantity surveying is dealt with in this book chiefly from the point of view of preparing a bill of quantities, but the ability to prepare quantities is also very necessary to the contractor for the compilation of tenders where quantities are not supplied, which is common in design and build and small contracts. A contractor may well produce quantities for an estimate including only the main items of work and not all of the items that would have been measured using *NRM2*. The descriptions would be shorter and usually the pricing is worked out alongside the quantities, thus avoiding abstracting and billing. The contractor's estimate is solely for internal pricing. If mistakes are made or short-cuts are taken which lead to errors, the contractor alone suffers. The contractor is free to adapt the general principles of measurement to the company's needs and the specifics of an individual project.

Nevertheless, the contractor's surveyor must be able to check a bill of quantities and measure variations on the basis of that bill. It is therefore essential that the contractor's surveyor should understand how the bill is prepared, and there should be no difficulty in adapting this knowledge to suit the somewhat different requirements when preparing quantities for a contractor's estimate.

Differences of custom

It must be understood that, as a good deal of the subject matter of this book is concerned with method and procedure, suggestions made must not be taken as invariable rules. Surveyors will have in many cases their own customs and methods of working which may differ from those given here, and which may be equally good or, in their view, better. The procedure advocated is put forward as being reasonable and based on practice. Furthermore, all rules must be adapted to suit any particular circumstances of the project in hand.

Method of study

It is advised in the first place to study Chapter 4 in order to grasp thoroughly the form in which dimensions are usually written, irrespective of whether a computer system will eventually be used to record dimensions in practice.

Knowledge of elementary building construction and simple mensuration and trigonometry is assumed; where knowledge is weak in these subjects, further study is recommended before proceeding further with measurement.

Chapter 5 explains some of the alternative systems that are used in practice, and Chapter 6 explains how girths and centre lines are calculated.

Chapter 7 contains notes on general procedure rules for taking-off which should be read before attempting to study actual examples of measurement, and to which subsequent referral may also be useful.

Chapters 8 to 19 represent the sections into which the taking-off of a small building might be divided, and these should be worked through one at a time. The principal applicable clauses of *NRM2* are referred to in each chapter and should be studied concurrently. After the chapter has been read, the examples should be worked through. It should be possible to follow every measurement by reference to the drawing.

The examples of taking-off in this book are small isolated parts of what could be the dimensions of a complete building and are not a connected series. When they have been mastered in their isolation, it will be much easier to see how they might be expanded and fitted together to make up the dimensions of a complete building.

Chapters 20 and 21 deal with preliminaries and bill preparation, which are more logically dealt with after taking-off, as this is often a separate process.

Examples

The examples are included to illustrate the methods of measurement of a small unit of a building. They assume that full specification clauses would be set out in preambles to the bill (see Chapter 21).

The dimensions that are set down in the dimension column when taking-off are given to the nearest two decimal places of a metre. Side casts (or waste calculations, as they are sometimes called) are used to calculate these dimensions, and are given in millimetres to ensure accuracy.

The examples in the chapters are presented in a traditional dimension format, this being considered the best system for a textbook and what the candidate will usually be faced with in the examination room. Abbreviations have been used for deductions where a description sufficient to recognise an item clearly is all that is required. The abbreviations used in the descriptions are listed in the 'Abbreviations' section at the beginning of this book.

Chapter 3
The Use of the RICS *New Rules of Measurement* (NRM)

Background

The *NRM* Project is arguably one of the most significant developments in quantity surveying practice since the publication of the SMM7 in 1988. The intention of the RICS with this publication is to create a set of common rules that provide a consistent approach to measurement through the various stages of a project, from initial cost estimate to detailed quantification of the construction work. Whilst the NRM is based on UK practice, it is nevertheless intended to have worldwide application.

This chapter therefore looks at the key features of the NRM in order to explain how it relates to the measurement covered in this text.

Historically, surveyors would approach the measurement of approximate quantities in different ways: for example, the area of external walls might be measured over windows and doors (i.e. gross measurement) by one surveyor, whilst another might deduct the area of the windows and doors (i.e. net measurement). The method of measurement would be closely related to the way in which the rates were to be applied. This variation in practice then resulted in an inconsistent approach to early estimates and cost planning, which then led to further problems when cost plans were used instead of bills of quantities as the basis of tender negotiations. Cost plans would then be analysed and used as benchmarks for further cost plans, thus creating an unreliable database and potentially inaccurate estimates. There was also a lack of continuity in cost data between cost plans and bills of quantities, making it almost impossible to reconcile the cost plan with the priced bill of quantities or pre-tender estimate.

Willis's Elements of Quantity Surveying, Twelfth Edition. Sandra Lee, William Trench and Andrew Willis.

Published 2014 by John Wiley & Sons, Ltd.

The NRM volumes

The RICS is in the process of publishing a suited set of documents in three volumes, for the measurement of building work from the early feasibility stage through to completion, handover and building occupation. The full NRM will comprise:

Volume 1 (NRM1) – Order of Cost Estimating and Elemental Cost Planning, covering:

- Estimating – RIBA Work Stage 1; Office of Government Commerce (OGC) Gateways 1 & 2
- Cost planning – Elements – RIBA Work Stages 2–4; OGC Gateways 3A, 3B & 3C

Volume 2 (NRM2) – Construction Quantities and Works Procurement, covering:

- RIBA Work Stages 5 & 6; OGC Gateway 3C (detailed measurement for tender documentation)

Volume 3 (NRM3) – Maintenance and Operation Cost Planning and Procurement, covering:

- RIBA Work Stage 7; OGC Gateways 4 & 5 (life cycle costing)

NRM1 was published in 2009 and updated in 2012 with the launch of NRM2. At the time of printing this book, there is no firm date given by the RICS for the publication of NRM3. The RIBA Outline Plan of Work is available from the RIBA website at www.architecture.com, and the UK Government Office of Government Commerce (OGC) Gateway methodology is detailed on their website at www.ogc.gov.uk.

Volume 1 – Order of cost estimating and cost planning for capital building works

This initial volume provides guidance on the quantification of building works for the purpose of preparing early cost estimates and cost plans. In addition, it gives direction on how to quantify other factors including preliminaries, overheads and profit, project team and design team fees, risk allowances and inflation.

Contents:

- Part 1: General. This places cost estimating and elemental cost planning in context and includes the RIBA Plan of Work and the OGC Gateway methodology definitions used in the rules.
- Part 2: Measurement rules for order of cost estimating. This *describes* the purpose and content of an order of cost estimate; defines its key constituents and explains how to prepare an order of cost estimate; and sets out the rules of measurement for the preparation of order of cost estimates using the floor area method, functional unit method and elemental method.
- Part 3: Measurement rules for cost planning. This *describes* the purpose of formal elemental cost plans, explains their key constituents and explains how to prepare an elemental cost plan.
- Part 4: Tabulated rules of measurement for elemental cost planning. This comprises the detailed tabulated rules of measurement for the preparation of approximate quantities for formal elemental cost plans.
- Appendices include definitions of gross internal floor area (GIFA), net internal area (NIA), measurement rules for elemental unit quantities (EUQ), information requirements for formal cost plans and templates for elemental cost plans.

For those not familiar with the RIBA works stages, these are included in Table 1.

The *structure* of an order of cost estimate is the same as that of Formal Cost Plan 1, thus enabling a consistent approach to be taken as further design details become available. (See Tables 2 and 3.) The *rules of measurement*, however, differ by taking into account the level of design detail available at each stage.

The different rules for the order of cost estimate as opposed to the elemental cost plan can be found by comparing Volume 1 Part 2 with Part 3.

Part 2, section 2.6 contains the measurement rules for building works which give two alternatives depending on the level of information available.

- The floor area method (*GIFA* × *appropriate cost/m*^{2})
- The functional unit (*functional unit, including all circulation space* × *appropriate rate per unit, i.e. the number of hospital beds or area of retail shops*).

Part 2, section 2.7 gives direction on the use of the elemental method at this early stage for elements where the information exists. The elements

Table 1 RIBA work stages (excluding Stage 0)

RIBA Stages		NRM Cost Stages	QS Action
1	Preparation and Brief	Order of Cost Estimate	Estimate developed from cost per functional unit through to cost per square metre and high-level elemental cost per GIFA
2	Concept Design	Formal Cost Plan 1	Develops order of cost estimate through to an elemental cost plan based on element unit quantities.
3	Developed Design	Formal Cost Plan 2	As more design information is released, greater detail can be included in the cost plan.
4	Technical Design	Formal Cost Plan 3 Pre-tender Estimate	Cost checks undertaken. Bills of quantities may be measured where required by the procurement approach being used.
5	Construction		Agreement of variations and payments for work completed Analysis of tender costs
6	Hand-over and Close-out		Settlement of any claims and final account
7	In Use		Evaluation of project undertaken Checks made against original cost targets.

Table 2 Structure of the order of cost estimate

Ref	Items	Currency
1	Building works	
2	Main contractor preliminaries	
3	Main contractor overheads and profit	
4	**Works cost estimate** Project/design team fees	
5	Other development/projects costs	
6	**Base estimate** Risk allowances: Design development risks Construction risks Employer change risks Employer other risks	
7	**Cost limit (excluding inflation)** Inflation: Tender inflation Construction inflation	
	Cost limit (including inflation)	

Table 3 Structure of Formal Cost Plan 1

Cost Centre	Group Element/Element	Cost/m^2 of GIFA	Total Cost of Element (Target Cost)
		Currency	Currency
BUILDING WORKS			
0	Facilitating works		
1	Substructure		
2	Superstructure		
3	Internal finishes		
4	Fittings, furnishings and equipment		
5	Services		
6	Complete buildings and building units		
7	Work to existing buildings		
8	External works		
SUB-TOTAL: FACILITATING AND BUILDING WORKS (A)			
9	Main contractor's preliminaries (B)		
	SUB-TOTAL: FACILITATING AND BUILDING WORKS (including main contractor's preliminaries) (C) [C = A + B]		
10	Main contractor's overheads and profit (D)		
TOTAL: BUILDING WORKS ESTIMATE (E) [E = C + D]			
	PROJECT AND DESIGN TEAM FEES AND OTHER DEVELOPMENT AND PROJECT COSTS		
11	Project and design team fees (F)		
12	Other development and project costs (G)		
TOTAL: PROJECT AND DESIGN TEAM FEES AND OTHER DEVELOPMENT AND PROJECT COSTS ESTIMATE (H) [H = F + G]			
	BASE COST ESTIMATE (I) [I = E + H]		
13	TOTAL: RISK ALLOWANCE ESTIMATE (J)		
	COST LIMIT (excluding inflation) (K) [K = I + J]		
14	TOTAL: INFLATION ALLOWANCE (L)		
15	COST LIMIT (excluding VAT assessment) (M) [M = K + L]		
16	VAT ASSESSMENT		Excluded normally

are those listed in Table 3. It is important to note that there can still be optional approaches as the design detail for some elements may not be available. For example, if the external wall details are not available, the cost of this element can be found by taking the GIFA × a rate per m^2 of the GIFA. If, however, the construction of the walls and windows is known, the cost would be the area of the walls (EUQ) multiplied by a suitable rate per m^2 (element unit rate, or EUR) of the walling. The method of measuring and the units are, then, those contained in Appendix E of Volume 1.

Part 3, section 3.11 contains the measurement rules for building works when preparing formal cost plans. A point of note is the statement that the degree of detail of measurement is to be related to the cost significance of the elements in the particular design. Where sufficient information is available, cost-significant items are to be measured by approximate quantities rather than the EUQ.

Part 3 has some duplication from Part 2 as there are common areas that need to be included in both the order of cost estimates and elemental cost plans. These include everything after the estimate for the building works. The use of NRM1 is therefore intended primarily to be restricted to the design stages before production or construction drawings are developed and prior to the preparation of detailed tender documents.

The potential benefits of using NRM1

NRM1 can be used as a tool for training and education to help develop cost planning skills and is a useful reminder of the items to include. It enables definitive lists of information to be provided to the design team so that they know what needs to be provided for each formal cost plan.

It includes, for defined terms such as building works, estimate and cost limits to provide clearer understanding and consistency. It has also provided a universal work breakdown structure (WBS).

Formal estimating stages have been introduced with clear frameworks to be followed, which should lead to more accurate cost estimates and cost plans. These now consider modern construction products and methods, including modular units, complete buildings and also sustainable construction.

Risk management has been promoted by the inclusion of risk allowances, and NRM1 dispenses with the use of the generic term *contingency*.

Inflation has also been considered in more detail with definitions of how to measure inflation and what items to apply this to.

The NRM has been issued as official RICS guidance, and this may create problems for any surveyor who does not follow the guidance should any of his or her methods be called into question. In the extreme, not following the guidance may also invalidate the surveyor's Professional Indemnity Insurance.

It is not the intention of this book to be a guide on how to apply the rules contained in the NRM1, as the focus here is on the use of NRM2 for the production of bills of quantities.

Introduction to NRM2

There is still a need for the preparation of bills of quantities following the rules of a standard method of measurement, where detailed information is provided as the basis for the traditional, fixed-price, lump-sum approach to procurement. NRM2 contains the rules for measurement to be used when preparing bills of quantities, and it is therefore used as the basis for explaining how to measure the examples included in this text. Table 4 gives the various work sections that are included in NRM2, and examples of measurement using these rules are included in Chapters 8–19.

NRM2 has a detailed section on how to code bills of quantities, but the link between elemental cost plans and bills in trade order can be seen only if bills are accurately coded when produced to enable sorting between the two different formats. Further information on bill preparation is provided in Chapter 21.

PDF copies of Volumes 1 and 2 of the NRM are available as a free download for RICS members from the RICS website to encourage all members to use the guidance. NRM1 is a sizeable volume but contains very good instruction on how to prepare early cost estimates and cost plans. It contains useful templates to guide those who do not have them provided by their company.

An interactive version is also available through the RICS knowledge portal iSurv, although this is available only by subscription.

Bound copies of NRM1 and NRM2 are available for purchase through RICS Books and other technical bookshops.

Little information is currently available about how the rules in NRM3 will be structured, and therefore there is no discussion on measurement for life cycle costing in this text.

Table 4 Extract from Section 3 of NRM2

	Work sections 2 to 41 comprise the rules of measurement for building components and items. They are as follows:
2	Off-site manufactured materials, components and buildings:
3	Demolitions;
4	Alterations, repairs and conservation;
5	Excavating and filling;
6	Ground remediation and soil stabilisation;
7	Piling;
8	Underpinning;
9	Diaphragm walls and embedded retaining walls;
10	Crib walls, gabions and reinforced earth;
11	In-situ concrete works;
12	Precast/composite concrete;
13	Precast concrete;
14	Masonry;
15	Structural metalwork;
16	Carpentry;
17	Sheet roof coverings;
18	Tile and slate roof and wall coverings;
19	Waterproofing;
20	Proprietary linings and partitions;
21	Cladding and covering;
22	General joinery;
23	Windows, screens and lights;
24	Doors, shutters and hatches;
25	Stairs, walkways and balustrades;
26	Metalwork;
27	Glazing;
28	Floor, wall, ceiling and roof finishings;
29	Decoration;

Table 4 (Cont'd)

	Work sections 2 to 41 comprise the rules of measurement for building components and items. They are as follows:
30	Suspended ceilings;
31	Insulation, fire stopping and fire protection;
32	Furniture, fittings and equipment;
33	Drainage above ground;
34	Drainage below ground;
35	Site works;
36	Fencing;
37	Soft landscaping;
38	Mechanical services;
39	Electrical services;
40	Transportation; and
41	Builder's work in connection with mechanical, electrical and transportation installations.

Comparison of SMM7 and NRM2

NRM2 is divided into three parts with supporting appendices:

- **Part 1: General** – This places the measurement for works procurement in context with the *RIBA Plan of Work* and the *OGC Gateway Process*, and it explains the symbols, abbreviations and definitions used in the rules.
- **Part 2: Rules for detailed measurement of building works** – This outlines the benefits of detailed measurement, describes the purpose and uses of NRM2, explains the function of bills of quantities, provides work breakdown structures for bills of quantities, defines the information required to enable the preparation of a bill of quantities, describes the key constituents of bills of quantities and explains how to prepare a bill of quantities. Considerable space is given to the codification of bills of quantities and the use of the bills for cost management.
- **Part 3: Tabulated rules of measurement for building works** – This comprises the majority of NRM2, being the tabulated rules for the

measurement and description of building works for the purpose of works procurement.

- **Appendices**:
  Appendix A: Guidance on the preparation of bills of quantities
  Appendix B: Template for preliminaries (main contract) pricing schedule (condensed)
  Appendix C: Template for preliminaries (main contract) pricing schedule (expanded)
  Appendix D: Template for a pricing summary for an elemental bill of quantities (condensed)
  Appendix E: Template for a pricing summary for an elemental bill of quantities (expanded)
  Appendix F: Templates for provisional sums, risks and credits
  Appendix G: Example of a work package breakdown structure

As can be seen from this brief list of contents, there is considerably more information given relating to the preparation of the bills of quantities than in SMM7. In the past, each quantity surveying practice produced their bills following their own preferences and historical development. This is the first time that the RICS has given guidance on coding and work breakdown structures in an attempt to align bills of quantities with the cost plans produced using NRM1.

Coding and work breakdown structures will not be covered here, and reference should be made to the detail contained in Part 2 of NRM2.

Comparison of SMM7 with NRM2

The following table has been prepared by reviewing the sections in SMM7 in order and identifying where the measurement rules are located and may have changed in NRM2. This is not a detailed comparison on a line-by-line basis but looks at the general measurement principles.

The framing of descriptions is also not as simple following these new rules as opposed to using SMM7, and care should be taken to ensure that each description adequately reflects the work to be priced. See Section 4 on phrasing descriptions.

One major change is in the layout of the document. There is now no reference to 'co-ordinated project information', and the lettering used previously has now been replaced with a numbering system. The comparison of the contents of each document appears in Table 5, and a comparison of the Part 3 rules in Table 6.

Table 5

<table>
<tr><th colspan="2">SMM7</th><th>NRM2</th></tr>
<tr><td colspan="2">General Rules</td><td rowspan="26">1 Preliminaries
2 Off-site manufactured materials, components and buildings
3 Demolitions
4 Alterations, repairs and conservation
5 Excavating and filling
6 Ground remediation and soil stabilisation
7 Piling
8 Underpinning
9 Diaphragm walls and embedded retaining walls
10 Crib walls, gabions and reinforced earth
11 In-situ concrete works
12 Precast/composite concrete
13 Precast concrete
14 Masonry
15 Structural metalwork
16 Carpentry
17 Sheet roof coverings
18 Tile and slate roof and wall coverings
19 Waterproofing
20 Proprietary linings and partitions
21 Cladding and covering
22 General joinery
23 Windows, screens and lights
24 Doors, shutters and hatches
25 Stairs, walkways and balustrades
26 Metalwork
27 Glazing
28 Floor, wall, ceiling and roof finishings
29 Decoration
30 Suspended ceilings
31 Insulation, fire stopping and fire protection
32 Furniture, fittings and equipment
33 Drainage above ground
34 Drainage below ground
35 Site works
36 Fencing
37 Soft landscaping
38 Mechanical services
39 Electrical services
40 Transportation
41 Builder's work in connection with mechanical, electrical and transportation installations (BWIC)
Appendices</td></tr>
<tr><td>A</td><td>Preliminaries/general conditions</td></tr>
<tr><td>C</td><td>Existing site/buildings/services</td></tr>
<tr><td>D</td><td>Groundwork</td></tr>
<tr><td>E</td><td>In-situ concrete/large precast concrete</td></tr>
<tr><td>F</td><td>Masonry</td></tr>
<tr><td>G</td><td>Structural carcassing metal/timber</td></tr>
<tr><td>H</td><td>Cladding/covering</td></tr>
<tr><td>J</td><td>Waterproofing</td></tr>
<tr><td>K</td><td>Linings/sheathing/dry partitioning</td></tr>
<tr><td>L</td><td>Windows/doors/stairs</td></tr>
<tr><td>M</td><td>Surface finishes</td></tr>
<tr><td>N</td><td>Furniture/fittings</td></tr>
<tr><td>P</td><td>Building fabric sundries</td></tr>
<tr><td>Q</td><td>Paving/planting/fencing/site furniture</td></tr>
<tr><td>R</td><td>Disposal systems</td></tr>
<tr><td>S</td><td>Piped supply systems</td></tr>
<tr><td>T</td><td>Mechanical heating/cooling/ refrigeration systems</td></tr>
<tr><td>U</td><td>Ventilation/air-conditioning systems</td></tr>
<tr><td>V</td><td>Electrical supply/power/lighting systems</td></tr>
<tr><td>W</td><td>Communications/security/controls systems</td></tr>
<tr><td>X</td><td>Transport systems</td></tr>
<tr><td>Y</td><td>Mechanical and electrical services measurement</td></tr>
<tr><td colspan="2">Additional rules – work to existing buildings</td></tr>
<tr><td colspan="2">Appendices</td></tr>
<tr><td colspan="2">Alphabetical Index</td><td></td></tr>
</table>

Table 6 Comparison of SMM7 and NRM2

Ref:	SMM7	Ref:	NRM2
A	Preliminaries		Similar provision, expanded and in greater detail added to give clear direction. Divided into two sections: Part 1 for the main contract and Part 2 for works packages if used.
		1	Main contract preliminaries
			Part A – Information and requirements Part B – Pricing schedule
		1	Preliminaries (works package contract) Part A – Information and requirements Part B – Pricing schedule
			Similar provision as SMM7 but expanded and with greater detail added to give clear direction.
B	Complete buildings	2	Offsite manufactured materials, components or buildings
			There were no details provided in SMM7, but now rules are included for prefabricated structures, and units (e.g. bathroom pods) in proprietary packages.
C	Demolitions and alterations	3	Demolitions
			Rules remain very similar; however, recycling provisions have been added.
		4	Alterations, repairs and conservation
			Rules covering conservation have been increased, and decontamination has been expanded; however, spot items have been removed.
D	Groundwork	5	Excavation and filling
			Generally, no change in principles

Table 6 (Cont'd)

Ref:	SMM7	Ref:	NRM2
D20.1	Site prep		Increased detail on site preparation items if used (e.g. boreholes and trial pits)
	Removing trees	5.2	Removing trees – the band widths have been varied.
D20.2	Excavating	5.6	Rules now simplified into either bulk excavation or foundation excavation, with depth ranges now only in 2 m stages.
		5.7	Earthwork support has been simplified in that it is now only measured where specifically called for in the contract documents (specification).
			There is no requirement to measure working space or compacting of surfaces.
		5.11	Filling is now either less than 500 mm or final thickness stated where over 500 mm thick.
		5.20	Cutting off the tops of piles has been included here instead of in the piling section.
		6	Ground remediation and soil stabilisation
			New section added
		7	Piling – no real change, although simplified by removing the separation of different types of piles, and cutting off tops has been moved to excavation.
		8	Underpinning – greatly simplified. Preliminary trenches and excavation measurements no longer required.
		9	Diaphragm walls – simplified, no excavation measured here, detail description stating thickness of wall

Ref:	SMM7	Ref:	NRM2
E	In-situ concrete/large precast	11	In-situ concrete
			Separated out precast from in-situ concrete, and radical change in the way concrete is measured
			New sections are:
			Mass concrete – cubic meters (m^3) any thickness
			Horizontal work – $m^3 \geq$ or $\leq$ 300 mm thick
			Sloping – m^3 > or < 300 mm thick
			vertical – m^3 > or < 300 mm thick
			Sundry – square meters (m^2) or m^3 > or < 300 mm thick
			Sprayed – m^2 thickness stated.
E20	Formwork	11.13	Formwork – some minor changes. If less than 500 mm wide, then measured by metre run.
			Mortises and holes have been removed from this section.
			Forming door openings has changed and these are now numbered.
E30	Reinforcement	11.33	Reinforcement – minor changes
			Spacers and chairs are now deemed included. Pre-tensioned members have now been added.
E40	Designed joints	11.38	Remains the same
E41	Worked finishes and cutting		Has been removed as a separate section
E42	Accessories	11.41	Accessories – no real change
E05.16	Grouting	11.42	Grouting and filling – new section added, now measurement chasing is by metre and grouting stanchion bases or filling holes by number

Table 6 (Cont'd)

Ref:	SMM7	Ref:	NRM2
E50/60	Precast concrete frames	12	Precast/composite concrete
			Now a section for composite precast units, measured m^2 for walls and floors
		13	Precast concrete
			Very similar; however, formwork and reinforcement are now deemed included.
F	Masonry	14	Masonry
			The natural and artificial stone rules are now all put under one section. All walls are measured along the centreline and assumed vertical unless stated.
			Skins of hollow walls added
			Separate facework items have now been removed, and there is limited use of the extra over provision.
G	Structural metalwork	15	Structural metalwork
			Major changes – weight classification now as follows: less than 25 kg, 25–50 kg, 50–100 kg or over 100 kg. Short lengths are now also included.
			Permanent formwork changed to profiled metal decking.
G20	Carpentry/timber framing/first fix	16	Carpentry
			Separated primary structural members (wall plates) from engineered or prefabricated members (roof trusses)
			All boarding, flooring and sheeting are now measured here; either ≤ 600 or > 600 mm wide.
G30	Metal profiled sheet decking	17	Sheet roof coverings

Ref:	SMM7	Ref:	NRM2
			Now covers, bituminous felt, plastic sheets, sheet metals and rigid boards.
			Some major changes in items deemed included in boundary work.
G31	Prefabricated timber decking		Removed
H	Cladding and covering		This has been split into numerous sections:
H10/12 /13	Patent glazing and structural glass (In SMM7, sections K11–21 are included together under H20.)		
H10, 12 and 13	Patent glazing/plastics/ structural glass	21	The range of claddings has all been included under Section 21.
H11	Curtain walling		Basic measurement principles remain the same.
H20, H21 and H92; K11, K12, K13, K14 and K15; H30, H31, H32, H33, H41 and H43	Sheet claddings, profiled sheet claddings and other claddings		The categories have been reduced; for example, there is no need to differentiate multi-tier roofs. However, widths less than 600 mm wide now need to be measured lineally.
21.1	Other chambers	21.8	Doors and openings are still measured as extra over the work in which they occur.
H10.4 and 5	Raking curved cutting	21.9	Boundary work – includes all cutting and trims to form the edge or intersection.
		21.1	Opening perimeters
			Edges of openings are measured here rather than as boundary work.
K20.1.1	Rigid sheet cladding	21.1	The narrow widths and isolated areas for other claddings are now no longer required to be measured.
H14	Concrete roof lights	13.2.3.4	Precast concrete roof lights are now measured with other precast concrete works. There is no change to measurement rules.
K20/21	Timber board flooring	16.4	Carpentry

Table 6 (Cont'd)

Ref:	SMM7	Ref:	NRM2
			Boarding is measured in metres if not exceeding 600 mm wide and in m^2 where over 600 mm wide.
H51, H52	Natural stone cladding / cast stone cladding	14.1	Masonry, natural stone walling and dressings, and cast stone walling and dressings
			Only measured in m^2 except for isolated features such as piers/ columns, arches etc.
			Perimeters and abutments are now measured as extra over, and more work is deemed included.
H51.5	Floors		This is not in the walling section and should now be referred to Section 28.2 (finishes).
H51.6–10	Staircase works		Refer to new Section 25.1, where staircases are now numbered.
H60	Plain roof tiling	18	Tile and slate roof and wall coverings
			Basic principles have not changed.
H60.11		41.27	Holes are now not measured with the roof but deemed included.
H70–H76	Metal sheet coverings etc.	17	Sheet roof coverings
		17.1	Now introduced width classification: < 500 mm wide (measured lineally) and > 500 mm wide (measured m^2).
			The allowances used in SMM7 when calculating areas have been removed. Now the information provided should identify all labours and dressings.
			Other principles remain the same.
J	Waterproofing	19	Waterproofing

Ref:	SMM7	Ref:	NRM2
J40			There is no separation of the item for tanking from coverings. Discretion is given to those drafting the description to include sufficient information to enable pricing.
J20.5	Skirtings	19.3	The width classifications have been reduced to now only < 500 mm wide or > 500 mm wide. Skirtings are no longer in band widths, but each is measured in metres stating the net girth on face.
J20.12	Internal angle fillets		Not measurable – deemed included
J41, J42 and J43	Built-up felt roofing	17	Sheet roof coverings
			Principles are similar, except width classification is now < 500 mm wide and > 500 mm wide.
			Girths at abutments are no longer applicable.
K	Linings, sheathing and dry partitioning	20	Proprietary linings and partitions
K10.1.1	Proprietary partitions measured in 300 mm stages; measured in metres.	20.1.1.2	Metal framed systems measured in m^2 in 1 m height stages; total length stated.
K32	Panel cubicles	22.16	Cubicle partitions – no change to measurement
K40	Demountable suspended ceilings	30.1	General principles remain the same. Fire barriers have now been introduced.
K41	Raised access floors	28.3	Measurement principles remain the same. Fire barriers within the void now need to be measured.
L	Windows, doors and stairs		
L10	Windows	23	Windows are in Section 23, doors in Section 24.

Table 6 (Cont'd)

Ref:	SMM7	Ref:	NRM2
			Bedding and pointing frames are no longer measured.
		23.8	Glazing supplied with windows and doors is now separated and no longer measured in m^2; panes are numbered with size given.
L20	Doors	24	Doors
			Measurement principles have not changed.
			Bedding and pointing frames are no longer measured.
P21	Ironmongery	24.16	Ironmongery
			Measurement principles have not changed.
L30 and Q41	Stairs and balustrades	25	Staircases
			No change to measurement principles, except the extra over items for ramps, wreaths etc. which are now deemed included.
L40 L41	General glazing Lead light glazing	27	Glazing
L40.1.1.1	Glass measured in m^2.	27.1.1.1	Glass now needs to have the panes numbered with the size stated. The category of special glass has been removed, but the type of glass needs to be stated.
L40.7–10	Engraving, etching etc. were measured in m^2 or design work numbered.	27.4	All additional features are now measured extra over and are numbered.
L40.13	Hacking out existing glass was measured in metres of the rebate length.	27.8	Removal of glass has changed to be measured by number, stating the size of panes.
L42	Infill panels	22.18	No change
M	Surface finishes	28	Floor, wall, ceiling and roof finishes

Ref:	SMM7	Ref:	NRM2
M10, M12, M13, M20 and M23; J10	Applied in-situ finishes	28.1	The width category has changed from < 300 mm to < 600 mm.
M10.16	Rounded angles and intersections		These are now deemed included.
M10.24	Accessories	28.25–35	The measurement principles remain the same; however, reinforcement, quilts are separated out from accessories. All accessories are now numbered. There is the opportunity to measure reinforcement either in m^2 or by the metre. No width categories are provided.
M10.26	Temporary supports		Deemed included; no longer measurable.
M21	Insulation with rendered finish	28.32	Insulation is measured separately from any rendered finish.
M22	Sprayed coatings		No specific rules separating this are included.
M30	Metal mesh lathing	28.30, 31	No specific section for this; it is measured as either metres or m^2, depending on the measurer's choice.
M31	Fibrous plaster		No separate rules for this.
M40/41/42/50/51	Tiling, parquet flooring and carpeting	28	No separation from in-situ finishes. Width category is now general at < 600 or > 600 mm.
			Description of carpeting should now include the underlay and edge grippers. Stair rods should still be measured and numbered as an accessory.
M52	Decorative papers	29.9	The areas that are measured have changed to being < 1.00 m^2 or > 1.00 m^2. Corners are now deemed included.
M60	Painting and decorating	29	Decoration
			General principles remain the same. The isolated surfaces have increased from –0.5 m^2 to 1.00 m^2.

Table 6 (Cont'd)

Ref:	SMM7	Ref:	NRM2
	Note Section M4.	29.1.2.1.1	Work to ceilings over 3.5 and not exceeding 5 m; and thereafter in 3 m stages and not 1.5 m.
M60.10	Coloured bands	38.16.6	These have been moved to the 'Services' section.
M61	Intumescent paint for fire protection	31.5	Now measured only in m^2
N	Furniture and equipment	32	Furniture, fittings and equipment
			Principles have not changed; however, the marking of positions, commissioning and connecting are deemed included as opposed to being measured, in accordance with Section Y of SMM7.
P	Building fabric sundries		
P10	Insulation, proofing and fire protection	31	Insulation, fire stopping and fire protection.
		31.1	Added ability to measure by metres as well as m^2, but no width categories given. Now includes for fire stops by the metre, and fire sleeves are numbered.
P11	Cavity insulation	31.6	No change
P20	Unframed items	22	General joinery
			General principles of measurement are unchanged; however, all ends, angles etc. are deemed included.
P20.8	Extra over hardwood		All the items are now deemed included.
P21	Ironmongery	22	General joinery
	This was ironmongery not supplied with windows or doors.	22.22	Now measured under Section 22; principles not changed.

Ref:	SMM7	Ref:	NRM2
P22	Sealant joints	22.19 and 20	Rules remain the same; however, there is now a separate item for raking out existing joints.
P30	Trenches/pipeways/pits for buried services	41.13	The external services have been separated from the internal ones.
P30.1	Excavating trenches	41.13	Service runs are now measured in 500 mm depth stages instead of 250 mm.
			There is no need to measure the bed and surround separately; it is now included in the description of the service run.
P30.9	Other chambers	41.17–22	The detailed measurement of the component parts of chambers has been removed, and these are now all numbered, stating the depth and giving a detailed description.
		41.27	A specific item has now been included for testing and commissioning of external services with provision for attendance and additional equipment.
P31	Holes/chases/covers/ supports for services	41	BWIC with M, E and transportation
P31.19	Cutting or forming holes	41.1	For electrical work, there is a removal of need to measure cutting and chasing by points. This is now one item per installation type.
			For other installations, there is no need to measure holes by their number – they are deemed included under the general builder's work item for the service.
P31.24 and 25	Ends of supports	41.6	These have been simplified, and the types of supports have been broadened to include pylons, poles, wall brackets, soffit hangers, stays and proprietary supports.

Table 6 (Cont'd)

Ref:	SMM7	Ref:	NRM2
P31.32	Work to existing structures	41.8 and 9	There is still a need to cut holes, chases and mortises into existing structures and lifting floor boards and chequer plates.
Q	Paving and fencing	35	Site works
			The principle of deducting for voids has increased from 0.5 m^2 to < 1.00 m^2.
Q10	Kerbs and edgings	35.1	No major change; accessories are now measured as extra over and are only numbered.
Q21	In-situ paving	35.6	No real change – measured in accordance with new Section 11.
Q22 and Q23	Coated macadam Gravel etc.	35.12, 35.13	Addition of band widths; now < 300 wide or > 300 mm
Q24 and Q25	Block and slab paving	35.14	Addition of band widths; now < 300 wide or > 300 mm.
Q26	Special surfacings	35.18–24	No change; however, the description for line marking has been broadened.
Q30 and Q31	Seeding and turfing; planting	37	Soft landscaping – no change
Q40	Fencing	36	Fencing – no change
R	Disposal systems		
R10/11	Above-ground drainage	33	Above-ground drainage
			The principles for the measurement have not changed.
R10.3, 4 and 5	Sockets, tappings and bosses	33.2	Ancillaries now include these items.
R10.7	Supports		These have been moved to BWIC. See above.
R10.15	Temporary operation		This has been removed from the measurement rules.
R12/13	Below-ground drainage	34	Below-ground drainage

Ref:	SMM7	Ref:	NRM2
			The rules for drainage have changed considerably; the excavation and beds and surrounds are not measured separately but are included in the drain run length.
R12.1–6	Excavation and beds and surrounds have all been removed.	34.1	Drain runs are now measured in average depth for the whole run in 500 mm depth stages.
R12.11–15	Chambers	34.6–11	All chambers are numbered, stating the depth with a detailed description.
		34.13 / 34.14	Sundries such as step irons and covers are still measured separately.
R12.19	O & M manuals		These are now covered under the 'Preliminaries' section.
X	Transport systems	40	Transportation
X1–12	Different systems	40.1	Greater detail has now been included for each system with the addition of separate items to be included for the following: offload and position equipment, assembly of component parts, free issue items, interface and connection to other systems.
X13	Marking holes	41.2	Moved to BWIC section
X14	Identification	40.3	This is now an item rather than numbered.
Y	Mechanical and electrical (M & E) services	38	Mechanical
Y10 and Y11	Pipelines and ancillaries	38.3	Pipework – there has been considerable change to the rules here.
Y10.1	Pipes	38.3	Generally, the pipe length is to include for all fittings and the like if insufficient design details are provided. The measure of the pipe now needs to identify the location of the installation.

Table 6 (Cont'd)

Ref:	SMM7	Ref:	NRM2
Y10.2.3	Fittings	38.4	There are two alternative rules which can be applied. The metre measure for the pipe can be deemed to include all fittings, or they can be separated and enumerated. The description of the fitting is to be precise, and the one end, two end approach has been dropped.
Y10.9	Pipe supports		Removed and included in pipe measure, or if special can be under BWIC.
Y11	Pipe sleeves	38.5	These are now included under ancillaries and again in the BWIC section.
Y20 and Y40	Equipment	38.1	Primary equipment
			Greater detail has now been included for each piece of equipment with the addition of separate items to be included for the following: offload and position equipment and assembly of component parts.
		38.2	Terminal equipment and fittings – separated out to identify equipment at the ends of pipe or duct runs.
Y20.6	Supports		These are now under the BWIC; not included as part of the equipment.
Y30	Air ductlines and ancillaries	38.6	This is by metre length; generally the length is to include all fittings and the like if insufficient design details are provided. The measure of the duct now needs to identify the location of the installation.
			There is an alternative of measuring the fittings separately.
Y30.7	Ducting sleeves		This has been removed, and the hole required is measured under BWIC.
Y50	Thermal insulation	38.9	Insulation and fire protection

Ref:	SMM7	Ref:	NRM2
			Generally, the principles have not changed. There is again the need to provide the location of the service.
			There is also the alternative here to include the insulation to fittings and the like within the measured length or separately.
		39	Electrical
Y60 and Y63	Conduit and cable trunking	39.3	Generally, containment is measured by length, including all fittings and ancillaries.
Y60.1–12		39.4	The rules have been greatly simplified; however, there is an alternative whereby the fittings can be enumerated separately.
			Forming holes or supports are covered under the BWIC section.
Y61, Y62 and Y80	High-voltage (HV) and low-voltage (LV) cables, busbar trunking, earthing and bonding	39.5	Cables are still measured by length, but the locations of the cables have been redefined.
			Terminations and glands are still enumerated, but supports and holes are now measured with the BWIC section.
Y62.7	Busbar trunking	39.9	Busbar
			This is by metre length; generally the length is to include for all fittings and the like. The measure of the busbar now needs to identify the location of the installation.
		39.10	There is an alternative of measuring the fittings separately.
Y80.13	Tapes	39.11	Tapes
			The length now includes all joints and test clamps; there is no alternative.

Table 6 (Cont'd)

Ref:	SMM7	Ref:	NRM2
			Electrodes and terminations are enumerated.
Y61.19	Cables in final circuits	39.7	Final circuits
			These remain as numbered circuits identifying the location of each circuit and the number and type of points.
		39.8	Modular wiring systems has been added. This is enumerated.
Y70 and Y71	HV switchgear, LV switchgear and distribution boards	39.1	Primary equipment
			Enumerated by system stating the location and with the addition of separate items to be included for the following: offload and position equipment and assembly of component parts.
			Supports, if not provided with the equipment, are measured under BWIC.
Y73 and Y74	Luminaires and lamps; accessories	39.2	Terminal equipment has been included and covers all the items in this section. They are again enumerated.
Y81	Testing and commissioning	39.15 and 16	Testing has been separated from commissioning, and further items have been added for system validation training and maintenance.
	Additional rules for work to existing buildings		The work to existing buildings has now been absorbed into the various work sections.

Chapter 4
Setting Down Dimensions

The development of computerised measurement and billing systems, each with its own structure for inputting dimensions and calling up descriptions, has made the more traditional procedures less common nowadays. However, it is only by understanding the basic principles of setting down dimensions and descriptions in traditional form, as detailed here and throughout this book, that one can then apply them to whatever measurement process is adopted.

Traditional dimension paper

The dimensions are measured from the drawings by the taker-off, using paper ruled as follows:

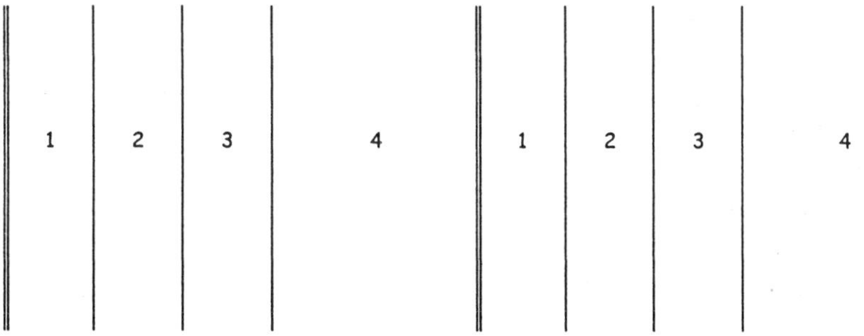

The columns (not, of course, normally numbered) have been numbered here for identification. Column 1 is called the *timesing column*, and its use will be described later. Column 2 is the *dimension column*, in which the measurements are set down as taken from the drawings. Column 3 is

Willis's Elements of Quantity Surveying, Twelfth Edition. Sandra Lee, William Trench and Andrew Willis.

Published 2014 by John Wiley & Sons, Ltd.

the *squaring column*, in which are set out the calculated volumes, areas and so on of the measurements in column 2. Column 4 is the *description column*, in which is written the description of the work to which the dimensions apply, and on the extreme right-hand side of which (known as *waste*) preliminary calculations and collections are made. There are two sets of columns in the width of a single A4 sheet. No written work should be carried across the central vertical division. There is usually a narrow binding margin (not shown here) on the left of the sheets.

Each dimension sheet should have the name and/or the number of the project written on or, better, stamped on. In addition, the title of the section being measured should be included, followed by a number, starting at 1 for each section. The unique numbering of the dimension sheets allows the surveyor to easily find an item at a later date. The examples measured in the following chapters have been written using only one-half of the sheet with the right-hand side being used for explanation.

Form of dimensions

Before going any further, it is necessary to understand the dimensions as set down by the taker-off. All dimensions are in one of five forms:

(1) Cubic measurements
(2) Square or superficial measurements
(3) Linear measurements
(4) Enumerated items
(5) Items.

These are expressed in the first three cases by setting down the measurements immediately under each other in the dimension column, with each separate item being divided from the next by a line, for example:

	3.00 2.00 4.00		indicating a cubic measurement 3.00 m long, 2.00 m wide and 4.00 m deep.
	3.00 2.00		indicating a superficial measurement 3.00 m long and 2.00 m wide.
	3.00		indicating a linear measurement of 3.00 m.

An item to be enumerated is usually indicated in one of the following ways:

	4		indicating four in number.
	Nr 4		
4/	1		

Occasionally, NRM2 requires the use of an item; this is a description without a measured quantity (e.g. testing the drainage system). The description may, if applicable, contain dimensions (e.g. temporary screens). This requirement is indicated as follows:

	Item		

There is no need to label dimensions *cube*, *super*, *lin* and so on as, if a rule is made always to draw a line under each measurement, it is obvious from the number of entries in the measurement under which category it comes. It is usual to set down the dimensions in the following order:

(1) Horizontal length
(2) Horizontal width or breadth
(3) Vertical depth or height.

Although the order will not affect the calculations of the cubic or square measurement, it is valuable in tracing measurements later if a consistent order is maintained. As will be explained in this book, an incorrect order in a description may even sometimes mislead an estimator in pricing.

The following is an extract from NRM2:

The rules for quantifying building components/items are as follows:

(1) *Measurement and billing*:
  (a) Measure work net as fixed in position unless otherwise stated.
  (b) Net quantity measured shall be deemed to include all additional material required for laps, joints, seams and the like, as well as any waste material.

(c) Curved work shall be measured on the centre line of the material unless otherwise stated.

(d) Dimensions shall be measured to the nearest 10 mm. 5 mm and over shall be regarded as 10 mm and less than 5 mm shall be disregarded.

(e) Except for quantities measured in tonnes (t), quantities shall be given to the nearest whole number. Quantities less than one unit shall be given as one unit. Quantities measured in tonnes (t) shall be given to two decimal places.

(2) *Voids*:

(a) Unless otherwise stated, minimum deductions for voids refer only to openings or wants within the boundaries of the measured work.

(b) Always deduct openings or wants at the boundaries of measured areas, irrespective of size.

(c) Do not measure separate items for widths not exceeding a stated limit where these widths are caused by voids.

Timesing

It often happens that, when the taker-off has written the dimension, it is found that there are several items having the same measurements. To indicate that the measurement is to be multiplied, it will be *timesed* as follows:

3/	3.00 2.00 4.00		indicating that the cubic measurement is to be multiplied by 3.
5/	3.00 2.00		indicating that the superficial measurement is to be multiplied by 5.

The timesing figure is kept in the first column and separated from the dimension by a diagonal stroke. An item timesed can be timesed again, with each multiplier multiplying everything to the right of it, as follows:

5/3/	3.00 2.00 4.00		indicating that the cubic measurement having been multiplied by 3, the result is to be multiplied by 5, i.e. the original measurement is multiplied by 15.
2/5/3/	3.00 2.00 4.00		indicating that the cubic measurement is to be multiplied by 30.

The timesing is done to a linear or enumerated item in just the same way as shown here.

Dotting on

In repeating a dimension, the taker-off may find that it cannot be multiplied but can be added. For instance, given that three items have been measured as follows:

3/	3.00 2.00 4.00		

To make the train of thought clearer, what is called dotting on can be used as follows:

indicating that the cubic measurement is to be multiplied by 3 + 2, i.e. by 5.

The dot is placed below the top figure to avoid any possible confusion with decimals, although these are usually avoided in timesing. Figures dotted on should be lower than the last, just as each one timesed is usually higher; this makes more space available than if they were all written in a horizontal line.

Timesing and dotting on can be combined, as follows:

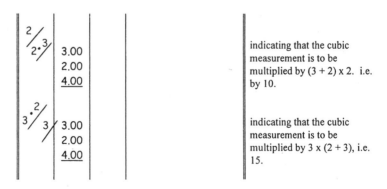

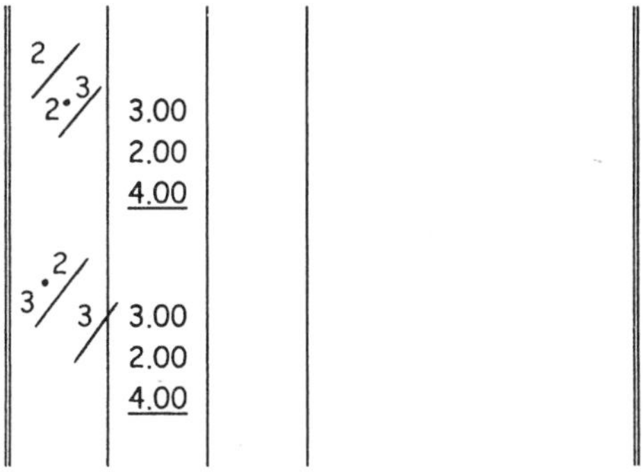

indicating that the cubic measurement is to be multiplied by (3 + 2) x 2. i.e. by 10.

indicating that the cubic measurement is to be multiplied by 3 x (2 + 3), i.e. 15.

Care must be taken when writing fractions in the timesing column that the line dividing the numerator from the denominator is horizontal, thereby avoiding any confusion with timesing.

Waste calculations

Except in very simple cases, dimensions should not be calculated mentally. Not only will the risk of error be reduced if the calculations are written down, because they will be checked, but also another person can readily see the origin of the dimension. These preliminary calculations, known as *waste calculations*, are made on the right-hand side of the description column. They must be written definitely and clearly, and not scribbled as if they were a calculation worked out on scrap paper. The term *waste* used for this part of the column might be thought to imply 'useless', but in fact it implies 'a means to an end'. Every effort must be made to commit to writing the train of thought of the taker-off. Waste calculations should be limited to those necessary for the clear setting down of the dimensions by the taker-off, and should not take the place of squaring.

An example of waste calculations is given in this chapter. These should not be written directly on the drawings. It is, however, necessary for dimensions to be traced from the drawings through any adjustments in the waste calculations to the figures used in the dimension column. They should preferably precede the description. An example of this is included along with the description for excavating a trench at the end of the 'Descriptions' section.

Alterations in dimensions

Where a dimension has been set down incorrectly and is to be altered, either it should be neatly crossed out and the new dimension written in, or the word *nil* should be written against it in the squaring column to indicate that it is cancelled. Where there are a number of measurements in the dimension column, care must be taken to indicate clearly how far the nil applies. This may be done as follows:

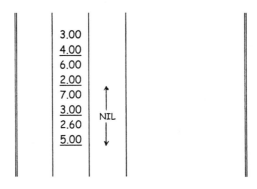

	3.00		
	4.00		

	6.00		
	2.00		

	7.00		
	3.00		

	2.60	↑ NIL ↓	
	5.00		

No attempt should be made to alter figures (e.g. a 2 into a 3, or a 3 into an 8). The figure may appear to have been altered satisfactorily, but it may look quite different to another person or when the dimensions are photocopied. Every figure must be absolutely clear, a page a little untidy but with unmistakable figures being far preferable to one where the figures give rise to uncertainties and consequent error. Deletions with correcting fluid should never be made as it is often of value to know what was written in the first instance. It is best to nil entirely and write out again any dimensions that are getting too confused by alterations; but great care is needed in copying dimensions, or mistakes may be made, it being particularly easy to miss copying the timesing. Therefore, where any dimensions are rewritten, they should be checked very carefully against the original.

The descriptions

The description of the item measured is written in the description column, on a level with its associated dimensions as follows (the waste calculation are also shown):

			34 000
			16 000
			2/50 000
			= 100 000
			- 4/215 860
			99 140
			15 050
			12 000
			13 500
			11 700
			4)52 250
			av = 13 060
			Underside of conc. - 12 050
			1 010
	99.14 1.00 1.01		Excavation, foundations not exceeding 2 m deep.

During the process of taking off, descriptions are often written in shorthand. The contents of the description are normally established with reference to the standard method of measurement. An extract from the concrete section of NRM2 has been reproduced in Fig. 1.

Item or work to be measured	Unit	Level 1	Level 2	Level 3	Notes, comments and glossary
Plain in-situ concrete **Reinforced in-situ concrete** **Fibre-reinforced in-situ concrete** **Sprayed in-situ concrete**					
1 Mass concrete	m^3	1 Any thickness	1 Infilling voids. 2 In-trench filling 3 In any other situation: details stated.	1 Poured on or against earth or unblinded hardcore.	1 Mass concrete is any unreinforced bulk concrete not measured elsewhere. 2 The volumes of each type of mass concrete work may be aggregated or given separately.
2 Horizontal work	m^3	1 ≤300 thick 2 >300 thick	1 In blinding. 2 In structures	1 Poured on or against earth or unblinded hardcore. 2 Reinforced >5%.	1 Horizontal work includes blinding, beds, foundations, pile caps, column base, ground beams, slabs, coffered and troughed slabs, landings, beams, attached beams, beam casings, shear heads, kerbs, copings, and upstands whose height is less than three times their width. 2 The volumes of each type of horizontal work may be aggregated or given separately. 3 Work laid in bays shall be so described, giving the average area of the bays.
3 Sloping work ≤15° 4 Sloping work >15°	m^3	1 ≤300 thick 2 >300 thick	1 In blinding 2 In structures 3 In staircases.	1 Poured on or against earth or unblinded hardcore. 2 Reinforced >5%.	1 Sloping work includes blinding, beds, slabs, steps and staircases, kerbs and copings. 2 Includes any attached beams, upstands, shear heads or similar. 3 The volumes of each type of sloping work may be aggregated or given separately. 4 Work laid in bays shall be so described, giving the average area of the bays.

Fig. 1
Source: From Section 11 of *NRM2*, by kind permission of the RICS.

A heading would be required identifying the kind and quality of the concrete; any tests of materials and finished work; measures to achieve water-tightness; limitations on the method, sequence, speed or size of pouring; and any methods of compaction and curing that might be specified. This can be found from the additional specification rules at the start of the concrete Section 11.

Typically a description for a specific item will be built up as follows:

- Location of the work: from the first column
- The second column identifies the unit of measure to be used.

- Depending on the item to be measured, further dimensions or descriptions required from columns headed Level 1, Level 2 and Level 3
- Any further information can be taken from the fifth column, where applicable.

A typical description of concrete in a suspended floor slab would be:

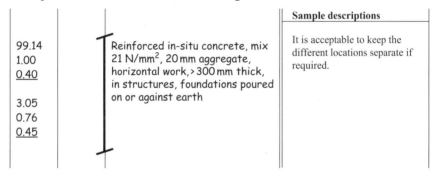

				Sample descriptions
	4.00 9.00 0.20		Reinforced in-situ concrete, mix $35 N/mm^2$, 20 mm aggregate, horizontal work, ≤ 300 mm thick, in structures	This item is measured in cubic metres; therefore, three dimensions are required in the dimensions column.

If two or more measurements are related to one description, it is normal practice for them to be bracketed together as follows:

				Sample descriptions
	99.14 1.00 0.40 3.05 0.76 0.45		Reinforced in-situ concrete, mix $21 N/mm^2$, 20 mm aggregate, horizontal work, > 300 mm thick, in structures, foundations poured on or against earth	It is acceptable to keep the different locations separate if required.

The bracket is placed on the outside of the squaring column. A clear indication is made of where the bracket ends, the bracket itself usually being a vertical line with a short cross mark to indicate top and bottom. The examples in the subsequent chapters are considered to be sufficiently clear without the need for the use of brackets.

This can be summarized as detailed in NRM2:

- Each work section of a bill of quantities shall begin with a heading and a description stating the nature and location of the work.
- Headings for groups of building components and items (i.e. components and sub-components) in a bill of quantities shall be read as part of the descriptions of the items to which the headings apply.
- Descriptions shall state the building components and items being measured (taken from the first column of the tabulated rules) and include all Level 1, 2 and 3 information (taken from the third, fourth and fifth columns, respectively) applicable to that item. Where applicable, the relevant information from column 5 shall be included in the description.

- Unless specifically stated otherwise in the bill of quantities or in these rules, descriptions for building components and items shall include the:
  (1) Type and quality of the material;
  (2) Critical dimension(s) of the material(s);
  (3) Method of fixing, installing or incorporating the goods or materials into the work where not at the discretion of the contractor; and
  (4) Nature or type of background.
- Where the nature or type of background is required to be identified, the building components and items description shall state one of the following:
  (1) Timber (the term includes all types of hard and soft building boards);
  (2) Plastics;
  (3) Masonry (the term includes brick, concrete, block, natural and reconstituted stone);
  (4) Metal, of any type;
  (5) Metal-faced timber or plastics; and
  (6) Vulnerable materials (the term includes glass, marble, mosaic, ceramics, tiled finishes, material finishes and the like).
- Dimensions given as part of the description shall be:
  (1) Stated in the sequence: length, width and height. Where ambiguity could arise, the dimensions shall be identified in the description.
  (2) The finished lengths, widths and heights specified or shown on the drawings with no allowance made for overlaps, scarcements, and the like.
- Thicknesses given as part of the description shall be the finished thickness of the material after compaction and shall exclude the thickness of adhesives and or bedding materials unless otherwise stated.
- The use of a hyphen (-) or the phrase 'to' between two dimensions in a description shall mean a range of dimensions exceeding the first dimension stated but not exceeding the second.
- Where the rules require work to be described as 'curved' with the radius stated, details shall be given of the curved work, including if concave or convex, if conical or spherical, if to more than one radius, and shall state the radius or radii. The radius shall be the mean radius measured to the centre line of the material unless otherwise stated.
- The information required by these rules may be given by a precise and unique cross-reference to another document (e.g. to a specification or to a catalogue).
- Where other components and sub-components are referred to in other documents (e.g. a specification states that vinyl sheet flooring is to be laid on a plywood lining), each component and or sub-component

shall be measured and described separately (i.e. both the vinyl sheet flooring and the plywood lining are to be measured as separate items).

- Notwithstanding the requirements above, separate components or sub-components may be combined to form single composite building components and items. In such cases, the description of the composite building components and items shall clearly state what is included and how each component and or sub-component is to be incorporated. Any component, sub-component or other element of the work not clearly included in the description shall be deemed not to be included as part of the composite building components and items.
- Unless specifically stated otherwise in the BQ or in these rules, each building component and item shall be deemed to include the following:
  (1) Labour and all costs in connection therewith;
  (2) Materials and goods together with all costs in connection therewith;
  (3) Assembling, installing, erecting, fixing or fitting materials or goods in position;
  (4) Plant and all costs in connection therewith;
  (5) Waste of goods or materials;
  (6) All rough and fair cutting, unless specially stated otherwise;
  (7) Establishment charges; and
  (8) Cost of compliance with all legislation in connection with the work measured, including health and safety and disposal of waste.

Anding-on

Where two or more descriptions are to be applied to one measurement, they are written as follows:

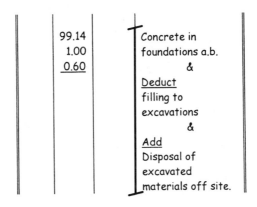

	99.14 1.00 0.60		Concrete in foundations a.b. & Deduct filling to excavations & Add Disposal of excavated materials off site.

In this case, each description is separated by '&' on a line by itself. Care must be taken when 'anding-on' in this way a superficial item with a linear item. It sometimes happens that it is very convenient to do so, but the distinction must be made quite clear so that the linear quantity is not used instead of the superficial. For example:

	5.33	5.33	25 x 100 mm wrot softwood rounded
	5.79	5.79	skirting plugged
			&
2/	7.00	14.00	Deduct
			Emulsion general
	4.50	4.50	surfaces, internally
		29.62	x 0.10 = 2.96 m²

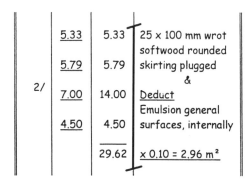

To measure the deduction of the emulsion paint in this way saves setting down all the dimensions again as superficial items (exactly the same lengths being used). Similarly, a superficial item might be marked to be multiplied by a third dimension to make a cube, for example:

	6.00		Cement and sand (1:3) screed bed, laid level or to falls ≤ 15° from horizontal
	4.50		
			&
			Reinforced in-situ concrete, mix 21 N/mm^2, 20 mm aggregate Horizontal work, ≤ 300 mm thick, in structures, bed
			x 0.15 = m^3

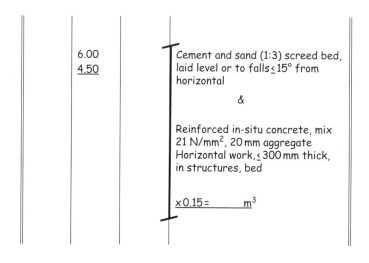

This approach is particularly useful for the measurement of roads, for example where you may have a large number of complex dimensions all relating to the various items associated with surfacing.

Deductions

Where a deduction is to be made, the description is preceded by *Deduct*, which is often abbreviated to *Ddt* but when carelessly written has been known to be mistaken for *Add*. It is, of course, important to make quite clear whether a measurement is to be added or deducted; and some surveyors always put the word 'Add' for any description immediately following a deduction, others only when a following addition is coupled to the deduction by '&', as in the example given at the end of this section. In taking-off on traditional paper, it is important that all deductions in a series of coupled descriptions are clearly marked 'Deduct', so that all doubt regarding whether any description is an add or deduct thereby is removed. To add emphasis to the words 'Deduct' and 'Add' used in this context, they are often underlined.

An example of deductions follows:

	1.00 2.10		Deduct One and half brick wall facework one side & Deduct Plaster brick walls & Deduct Emulsion plaster general surfaces internally
	1.00 1.20		Deduct Gloss paint plaster general surfaces internally & Add Two coats emulsion plaster general surfaces internally

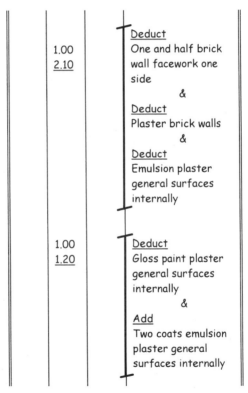

Spacing of dimensions

One of the commonest faults found with the taking-off among those new to measurement is the crowded state of their dimensions. All measurements and descriptions should be spaced well apart, so that it is quite clear where one begins and the other ends. It is not unusual for a taker-off to realise, after writing down the measurements, that some item has been overlooked and should be inserted in its proper place. If the dimensions are well spaced out, it can be squeezed in, but otherwise it will have to be inserted elsewhere and cross-references made, which only complicate the work. The use of a few extra sheets of paper will be found well worthwhile.

Accuracy

There may be a temptation, especially to those not sure of their ground, to take measurements on the full side, and so cover themselves against possible claims for deficiency. Such practice is to be deprecated, as it will be realised that, if measurements were on an average 2½%, or perhaps 5%, over-measured, a very substantial amount would be added to the tender. Where figured dimensions are the basis the measurements taken should be exact, but where they are scaled there is some excuse for measurements being slightly full, as there may be some shrinkage of the paper and the lines on the drawing are not usually drawn more precisely to scale than, say, 50 mm on a 1:100 scale plan with any certainty. It will be seen, therefore, that the care with which the architect's drawings are prepared may affect the quantities, and that measurement from fully figured drawings must necessarily be more accurate. Furthermore, the quantity surveyor owes a duty to the client to prepare an accurate bill of quantities, and it could be regarded as negligent not only to under-measure but also to over-measure.

When working from figured drawings, it may be found that three places of decimals are in some cases indicated. In setting down the dimensions, the figures should normally be taken to the nearest two places of decimals, the excess or loss in measurement thereby obtained being usually very slight in proportion to the whole, particularly when the total bill quantity is usually given to the nearest metre. Three places may, however, be desirable when setting down waste calculations.

In short, the surveyor must aim at giving in the bill a representation as accurate as possible under the circumstances of the building work in question. There is, however, a sensible limit to the degree of accuracy to which one should measure; more experienced takers-off often take the value of the item into account when deciding on the degree of accuracy required.

Numbering the dimension sheets

All dimension sheets should be numbered as soon as possible. For those in traditional form, some surveyors number each column, others each page, this being a matter of individual preference. If the numbering is done at the earliest opportunity, it will minimise the risk of a page being mislaid and overlooked. If sheets are inserted after numbering, they can be numbered with the last number and a suffix a, b, c and the like, but it is essential that, if sheets numbered, say, 31a, 31b, 31c and 31d are inserted, a note should be put against the number 31 stating that '31a–d follow'; otherwise, the inserted sheets might be lost and not missed. A similar note should be made if gaps occur in numbering.

In any case, it will be found of value if the taker-off numbers the pages at the top, according to the section they represent (e.g. floors 1, roofs 6, plumbing 9 etc.), irrespective of any other system of numbering. These pages can be kept in order awaiting the main sheet numbering, which may follow on from another section. It is further of value if on opening the dimensions at any page, one has an immediate indication of which section is measured.

Cross-references

The taker-off should try to ensure that dimensions are clear to others, as it is quite possible that, when variations on the contract have to be adjusted, someone else will be entrusted with the work and will have to find their way around the dimensions. Cross-referencing between the drawings and the dimensions is essential for ease and speed in locating why particular dimensions have been used. It has been known for projects to be postponed for a year or two after tenders were received, and even the taker-off will then need some references and notes to refresh the memory. The dimensions may also need to be referred to when preparing the final account, and again clarity is important.

Clearness of the dimensions

Besides the use of cross-references, a good deal can be done to make the dimensions clear by the manner in which they are set down. It has already been pointed out that a regular order of length, width and depth (or height) should be maintained in writing down the dimensions; even when it may be difficult to determine which is length or width, a consistent order should be kept. In measuring areas of floor finishes, for instance, the dimensions horizontal on the plan could be put first, followed by the vertical ones. Calculations should be made as waste on the dimension paper and not mentally, and timesing should be done consistently.

For instance, in measuring six doors each with four squares of glass, all timesed for two floors, the dimensions should be timesed as follows:

2/6/	1		50 mm wrot softwood four-panel door etc.
2/6/2/	0.90 1.95		One undercoat and two top coats, with oil-based paint on the general surfaces irrespective of the pane size.

with the last item not being written as:

6/2/2/	0.90 1.95		One undercoat and two top coats of oil-based paint on general surfaces irrespective of pane size.

When timesing becomes complicated, it will help considerably in tracing items if the method of timesing is consistent. In the example given here, the outer timesing represents the floors, and the next the number of doors; it could be confusing if the order is reversed in the middle of a series. The use of coloured pens in timesing or dotting on to represent different floors or sections of the work will be found to help considerably in tracing dimensions later.

Headings

The use of headings in the dimensions will further aid future reference. Apart from the taking-off section heading already suggested for each page, subheadings or *signposts* should be clearly written wherever possible, and these will stand out if underlined. The sequence of measurement in the section will then easily be followed by a glance at the subheadings.

Notes

The making of notes by the taker-off on the dimensions is of the utmost value. Such notes are usually for reference before the taking-off is finished (e.g. notes of items to measure or queries to be settled).

This type of note should be written at the right-hand side of the description column, and it is best separated by a line or bracket to prevent confusion with descriptions. There is often a column provided on data entry sheets for notes or comments to be added.

The type of note, made for reference before the taking-off is finished, is necessary when perhaps some point must be referred to the architect or engineer, or for other reasons something cannot be finally measured. As mentioned, a list of such queries should be compiled. *To take* notes are entered in dimensions when a taker-off dealing with one section feels that an item, although arguably within the work being measured, is better taken with another section (e.g. tile splashbacks to sinks taken with the plumbing services rather than with the finishes or vice versa). These *to take* notes should be written clearly in the dimensions and collected together before the bill is prepared, and a check should be made to ensure that all items listed have been measured. Memoranda, too, should be made at the end of the day's work of anything unfinished which, the train of thought being broken, might be forgotten. It is much safer to make notes of such matters on the dimensions than to trust them to memory, and these notes enable the work to be carried on in the case of unexpected absence. Such notes as these are sometimes written in pencil to be erased when dealt with, or they may be written in ink across the description column in such a way that they cannot be missed.

Insertion of items

It quite often happens that the taker-off must go back to the work and make alterations or insert additional items, either because items have been forgotten or because revised details have been received, as described in this chapter. Such alterations and additions should wherever possible be made in the proper place in the dimensions, so that when the dimensions are referred to at a later stage, everything can be found where expected. Once more, the importance of plenty of space must be emphasised, and on work of any size it will be found of advantage to start each section of the taking-off on a new sheet, leaving any odd blank columns for later use if necessary. Further, if the dimensions are kept in subsections or groups with a definite gap between, these gaps will also be found of use for the insertion in their proper place of any dimensions as an afterthought. If it is impossible to insert an item or group of items in its proper place, a place must be found for it elsewhere and proper cross-references made in both places, as described in this chapter.

Squaring the dimensions

A check is made of the waste calculations and their correct transfer to the dimensions. The dimensions are then calculated or *squared*, and where bracketed together they are totalled, subtracting any deductions that follow immediately. The *squarings* and casts are then checked and ticked.

Scheduling dimensions

The need for setting down dimensions on dimension paper in the traditional format can be quite time consuming for some items, and it can be beneficial to use a schedule to collect dimensions for repetitive items, such as steel beams or pipework. Take-off schedules may be prepared on A3 paper or a computer spreadsheet, identifying the item to be measured and referencing the schedule sheet. A location column is normal down the left-hand side, with perhaps varying sizes across the page. The right-hand side could be used for fittings, accessories, girths of beams or other notes.

The small extract from a schedule shown in Fig. 2 has been used to measure concrete beams on an Excel spreadsheet, and it is ideal when a building is set out to a grid pattern and beams are repeated over a number of floors. A collection of lengths at the end of the schedule would then be transferred to the dimension paper or entered directly into a computer package. In Fig. 2, the length of a beam on grid line 1 A–E is 7.90 m. This is also repeated on grid line 2; therefore, the figure is multiplied by 2 to give a total length of 15.80 m. Errors in working out waste calculations or extending dimensions are avoided when using a spreadsheet in this way. Care is still required to measure the lengths of the beams and to avoid any data entry errors.

Beams	Reinforced concrete beam sizes							
	400 × 500 mm		350 × 400 mm		350 × 300 mm		200 × 300 mm	
1st floor								
1 & 2, A–E	2/7.90	15.800						
A & E 1–2	2/5.10	10.200						
A & E 2–3			2/2.975	5.950				
3 A & E				7.200				
A & E 3–6					2/8.825	17.650		
5 & 6, A–E					2/7.20	14.400		
c, 3–5						7.150		
4, B–E						5.650		
B, 3–5								7.150
Lengths		26.000		13.150		44.850		7.150

Fig. 2

Schedules are often used to measure steelwork, drain pipe runs, pipework, ductwork and cables, but any item that is very repetitive could be measured in this way. The examples used in this textbook are not complex enough to warrant the use of schedules for measurement, but further explanation of a drainage schedule has been included in Chapter 17.

Chapter 5
Alternative Systems

This book concentrates on the traditional taking-off in order to explain the basic principles which can then be applied to alternative measurement and billing systems.

This chapter briefly outlines the alternative systems that have been developed and adopted in order to achieve standardisation, increase the efficiency of bill production and eliminate time-consuming data processing activities.

Computerisation of the measurements and bill preparation processes has been evolving over a period of time based on the use of standardised descriptions and advances in computer hardware and software.

Standardisation

The art of taking off is to a large extent based on developing a systematic approach, having a sound knowledge of building construction and acquiring the mathematical skill to calculate and measure dimensions and the ability to write clear descriptions. Of these, probably the latter is the most difficult to master because it has to be built up from long experience, especially in estimating and dealing with errors and claims arising from inadequate descriptions. The taker-off has to ensure that descriptions comply with the requirements of the Standard Method of Measurement (SMM) and refer properly to British Standards, and often has to frame them under pressure of time from complex and sometimes incomplete drawings and specifications. It is no wonder, therefore, that occasionally errors and omissions occur in descriptions, and it is not

Willis's Elements of Quantity Surveying, Twelfth Edition. Sandra Lee, William Trench and Andrew Willis.

Published 2014 by John Wiley & Sons, Ltd.

unknown for there to be diversity between different bills from the same office or even between different sections within the same bill, particularly in elemental formats. Furthermore, both personality and experience influence the taker-off's approach to compiling descriptions; and brevity, verbosity and English literacy all have their effect.

In former years, an assistant became proficient in framing descriptions by experience gained in a working-up section preparing abstracts and bills. With the introduction of computerised systems coupled with longer full-time education, this form of experience has been lost and alternatives have had to be established.

Standardisation was achieved either by compiling a standard bill or library within an individual organisation, including all the descriptions likely to be encountered in the type of work usually dealt with, or by using one of the published standard libraries.

Standard libraries

Most libraries rely on the fact that phrases within descriptions are frequently repeated, whereas the complete description is often peculiar to one situation. Therefore, through splitting descriptions into phrases and allocating the phrases to various levels, full descriptions can be built up by selecting phrases at each level, some being essential and others optional. Considerable saving in bulk is achieved as phrases that are frequently used are listed once only.

The levels of phrase in a standard library could be as follows:

- Level 1: main work section
- Level 2: subsidiary classification
- Level 3:main specification in heading
- Level 4: description of item
- Level 5: size and number
- Levels 6 and 7: written short items (extra over items).

By selecting obligatory and optional headings and phrases from each level, combinations are built up to complete the description. The advantages of such a system are numerous; some of them are summarised in the following list:

- Descriptions are standardised and consistent.
- It provides an *aide-mémoire* for the taker-off.
- Billing can be carried out by less trained staff.

- Bill editing is greatly reduced.
- Consistency between bills aids price comparison of items.
- The simplest possible wording, almost in note form, avoids ambiguities.
- The use of 'ditto' (often causing doubt) is avoided.
- It complies with the SMM.
- Consistency between bills from various sources aids the estimator.

This list would appear to give standard libraries an overwhelming advantage. There is, however, a danger that the taker-off will apply a readily available standard description to the item being measured even though it does not fit exactly, rather than compiling carefully a proper description; such an item is commonly called a *rogue item*. Furthermore, there is no doubt that if the taker-off has to keep referring to bulky documents, the train of thought is interrupted and speed is reduced.

Computerised bill production

With rapid advances in computer technology, and as computer hardware and software applications have become comparatively less expensive, many firms have adopted one of the many computer systems now available. They have become widely used not only in bill production but also for assistance in carrying outmost other quantity surveying functions. It is desirable that the operation of these other functions be carried out without having to re-enter the data stored for bill production, and therefore database systems are used.

A wide variety of bill of quantities production systems are now available, which have varying degrees of sophistication.

Although paper-based taking off can be manually entered into the systems, the majority now available have the facility to input dimensions directly by using either the keyboard or a digitiser. A digitiser is an electronically sensitive drawing board from which dimensions may be electronically scaled from the drawings into the system. Lengths, perimeters and areas of both simple and complex shapes are calculated automatically.

All systems are based on having a standard library of descriptions built up from library text displayed on the screen. The problem facing the industry is that there are a number of different standard libraries used by each supplier of bill preparation software, and often each company or practice that uses the software creates its own library from scratch. Some companies use the library provided with the software as a starting point and then add their own rogue items that they frequently use. When a quantity surveyor moves between companies, he or she will find the basic principles of the various systems similar, but there will be a need for re-training on the different software packages to ensure that the full features of the system are understood and utilised.

When a particular description is not available from the standard library, there is a facility to create rogue items. The number of rogues is likely to be more in a complex building project than in a simple one It is therefore the ease with which creating and sorting rogues can be carried out that determines how efficient a particular system is.

An extract from a typical library of descriptions is shown in Fig. 3. This figure shows a sample of the various levels of the description that are available, along with a coding system. The main heading might only be used once in a bill of quantities, but the common brickwork may well be required in both 102.5 mm and 215 mm thicknesses. If you look at the section in SMM7 for brickwork and blockwork you will see an extensive list of potential items that would need to be included in a library database.

Correctly coding items measured on paper for subsequent entry into a measurement package is of paramount importance, and there is always the danger of mis-coding. A record is kept, however, and can be physically checked.

Code	Description	Unit
F10	F10 BRICK / BLOCK WALLING	
F1001	COMMON BRICKWORK	
F1003	FACING BRICKWORK	
F10 - - 001	Walls	
F10 - - 003	Isolated piers	
F10 - - - - - 01	102.5 mm thick	m^2
F10 - - - - - 02	215 mm thick	m^2
F10 - - - - - 03	.... mm thick	m^2
F10 - - - - - - - 01	battering	
F10 - - - - - - - 02	tapering one side	
F10 - - - - - - - 03	tapering both sides	

Fig. 3

In contrast, some packages allow you to search through the library of descriptions to chose the item to include in your take-off, and it is less likely an error will be made at this stage. Dimensions can then be entered with the relevant description along with any waste calculations or side notes. A continual print of the data entry is useful so that you have a record of the logic used for the taking-off. This can prove very useful at the post-contract stage when researching how the bill was measured so that variations can be assessed.

The order of taking-off for direct computer entry may differ from that used in the traditional approach. Utilising the 'anding-on' facility, or as some suppliers of software call it 'carry forward', can be extremely useful to save re-entering a long stream of dimensions. Before you start a section of measurement, think carefully about how you want to order your work.

Common features with software packages are the facility to utilise data from previous bills of quantities, to calculate the weight of steel and reinforcement automatically and to display rules for calculating volumes. Systems can contain standard libraries for different standard methods of measurement and, being database systems, they have a multiple sort facility and can produce different bill formats as required. They can also sort, display and print data to different levels of detail. Draft bills can be produced for editing, and in many ways the editing process is more important than when using manual methods of bill production.

The development of bill production systems is ongoing, and these systems are continually being upgraded. Many systems already have the facility to generate bill items directly from computer-aided design (CAD) data or other schedule data. As such fully computerised systems continue to be developed and become more widespread, the quantity surveyor will concentrate more on interpreting the data produced, ensuring their completeness and utilising them not only for tender documentation production but also for cost control and post-contract administration purposes.

Electronic data interchange (EDI) is also becoming more widely used: the bill is issued in electronic form to the contractors, who can then use it to price their tender for the work. The advantages of this are that it speeds up the transmission and receipt of information, reduces the paperwork involved and eliminates the wasteful repetition of rekeying information into a separate system (which is time consuming and error prone).

E-tendering

The process of e-tendering is where the tender for a project is managed entirely electronically over the internet.

The RICS have published a guidance note for their members on what a system should involve and how it should be managed. This puts forward the proposition that the primary strategic purpose behind e-tendering is the desire to drive down costs. There are, however, other benefits that would support its wider use, such as the potential for aggregation of buying power, the elimination of waste, the simplification of the process and a fairer assessment between tenders.

Without the need for the production of multiple copies of paper-based information, there is a less environmentally demanding and more sustainable tender process.

The impact of e-tendering on measurement is that bills of quantities need to be produced electronically, and it should be possible for them to be converted into a commonly accepted format by contractors for pricing.

The RICS see the barriers to successful implementation of e-tendering as being a lack of agreed standards, little or no impartial advice, and the potential legal and technical traps. Hence there is a very slow move towards its acceptance.

Space is not available here to discuss the process in detail, and those wishing to obtain further information can do so through the RICS website.

Site dimension books

When surveyors are visiting a site it helps to have a notebook that is already ruled up for measurement. Typically these notebooks are just slightly smaller than A5, so that they are easily held, and have a hard cover to protect them throughout the life of a project.

The pages are ruled so that dimensions can be set down in the traditional format as shown below:

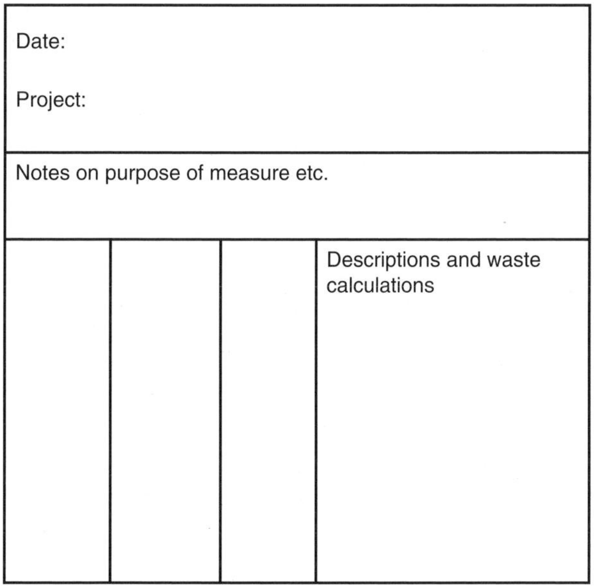

Date:
Project:
Notes on purpose of measure etc.
Descriptions and waste calculations

As site visits may not occur every day, it is important to note the date that a measurement was made and the purpose of the measurement. It may be to record details of a variation, or a measure of work for interim payment. Sketches, drawing references and location notes within the building to identify precisely where the work is located should also be made.

The site dimension book should be numbered for the project, as there may well be quite a few books over the life of the project, and the name of the surveyor recorded at the front. These books should then be filed when the project is complete so that there is a record should there be a dispute during agreement of the final account.

Estimating paper

Slightly different paper again is used when an approximate estimate is being prepared so that the dimensions and costing can occur on the same page. This type of paper can be purchased either loose by the ream or in A4 pads. A typically page would be follows:.

	1	2	3		4	5	6	Currency
				Waste calculations				
				Description				
				Cost build up				

Columns 1 to 3 are the traditional dimension columns, column 4 is the total quantity measured, column 5 is the unit of measure and column 6 is the rate to be used to price the item.

Chapter 6
Preliminary Calculations

Mathematical knowledge

It is assumed that the reader is acquainted with geometry and trigonometry, knowledge of which, as of building construction, is an essential pre-requisite to the study of quantity surveying. It is rare that more detailed knowledge is required than the properties of the rectangle, triangle and circle. Less known formulae are detailed in the Appendix and can be looked up in an appropriate mathematical reference book. The properties of the rectangle, triangle and circle should, however, be thoroughly understood; if a student is not acquainted with them and with elementary trigonometry, it is recommended to study these before going any further.

This chapter shows some examples of how the theoretical knowledge of mensuration is applied to building work. Wherever possible, lengths should be found by calculation from figured dimensions on the drawings,

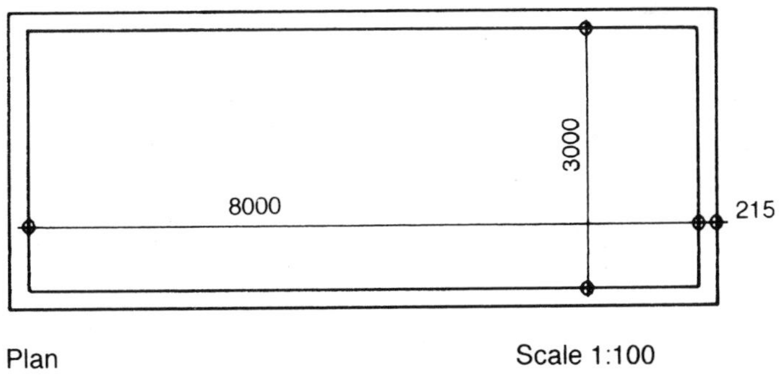

Fig. 4

Willis's Elements of Quantity Surveying, Twelfth Edition. Sandra Lee, William Trench and Andrew Willis.

Published 2014 by John Wiley & Sons, Ltd.

rather than by scaling. Where scaling has to be used, a check should be made to ensure that other figured dimensions are accurate, as some reproduction methods affect the scale. If using CAD drawings, a check on the scale should be made and the drawings calibrated.

Perimeter of buildings

Figured dimensions may be given internally or externally, and either can be worked from. Fig. 4 shows a plain rectangular building with one brick wall 215 mm thick.

	8.000
	3.000
	2/ 11.000
internal girth	22.000

The length of the perimeter of the external face of the walls, which may be required for the measurement of external rendering, can be calculated by adding twice the thickness of the wall at each corner.

	22.000
4/2/ 0.215	1.720
external girth	23.720

This can be confirmed by adding the thickness of the wall to each internal dimension first.

	3.000	8.000
2/ 0.215	0.430	0.430
	3.430	8.430
		3.430
		2/ 11.860
		23.720

It will be seen that twice the thickness of the wall has been added for each corner.

Centre line of the wall

NRM2 requires a brick or block wall to be measured net in square metres along its centre line, stating the thickness of the wall. The measurement of items such as foundation trenches, concrete footings and forming cavities all use the same centre line for the length when calculating their quantities.

As it is the net measurement that is required, there should be no duplication at the corners of a building. By looking again at Fig. 4, it can be seen that two walls of 8.43 m and two walls of 3.00 m are required, giving a net length around the centre line of 22.86 m. This should be apparent just by looking at the plan; however, most plans are not a simple rectangular shape, and waste calculations will be necessary to establish the required centre line. The detail of a corner is shown in Fig. 5, showing that the centre line requires either the thickness of the wall to be added or deducted from the internal or external dimensions, respectively.

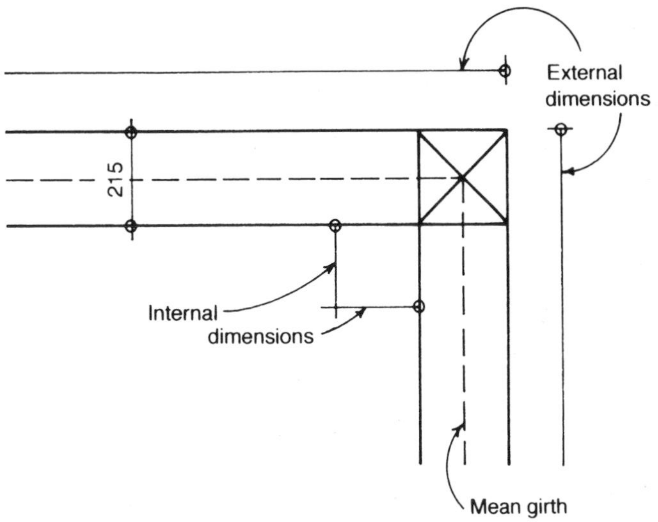

Fig. 5

A straightforward way of calculating the centre line measurement or mean girth from the internal dimensions is as follows:

$$
\begin{array}{rr}
\text{a.b.} & 22\,000 \\
\text{plus } 4/2/\frac{1}{2}/215 = & 860 \\
\hline
 & 22\,860 \\
\hline
\end{array}
$$

It will be seen that twice of half the wall thickness is added to the internal perimeter for each external corner, that is, $4 \times 2 \times$ the distance moved (half the thickness of the wall).

The whole process can, of course, be reversed. If the external perimeter is taken instead of the internal, by deducting instead of adding, the mean length of the wall is obtained as follows:

	23 720
Less $4/2/\frac{1}{2}/215 =$	860
	22 860

By becoming familiar with the centre line calculation, more complex shaped buildings can be measured quickly and repeated use made of the same dimension. It is therefore important that this dimension is correct as a simple error can be repeated through many items in a bill of quantities.

If the shape of the building is slightly more complicated, as shown in Fig. 6, the same principle may be applied to the calculations. For example, the centre line or mean girth measurement is:

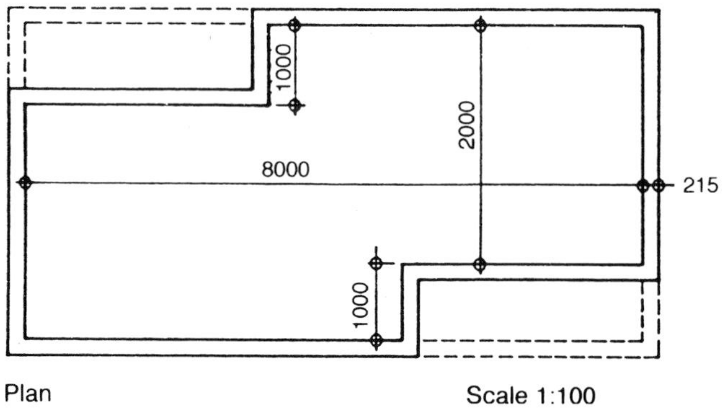

Fig. 6

	2 000
	1 000
	3 000
	8 000
	2/11 000
Inside face	22 000
plus $4/2/\frac{1}{2}/215 =$	860
Mean girth	22 860

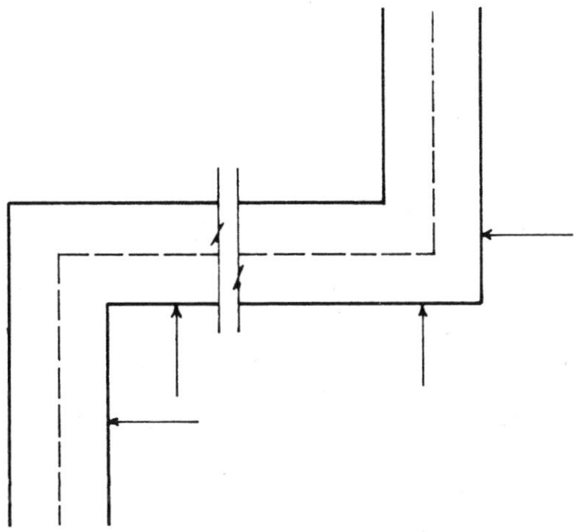

Fig. 7

Where the wall breaks back (enlarged in Fig. 7), the internal and external angles balance each other. As before, if the inside face has been measured, the external angle needs the thickness of the wall to be added to give the length on the centre line, whereas in the case of the internal angle the thickness of the wall must be deducted. In fact, the perimeter of the building is the same as if the corners were as shown dotted in Fig. 6. In short, to arrive at the mean girth, the thickness of the wall must be added for every external angle in excess of the number of internal angles.

This collection of the perimeter of walls being of great importance, one further and more complicated example is given in Fig. 8. This time, the calculations are made from the external figured dimensions instead of from the internal ones.

When smaller dimensions are given, you could check that their total equals the overall dimensions given, for example:

2 000	3 600		950
2 750	900		950
2 000	2 250		3 650
		1 200	5 550
6 750	6 750	– 750	+ 450
			6 000

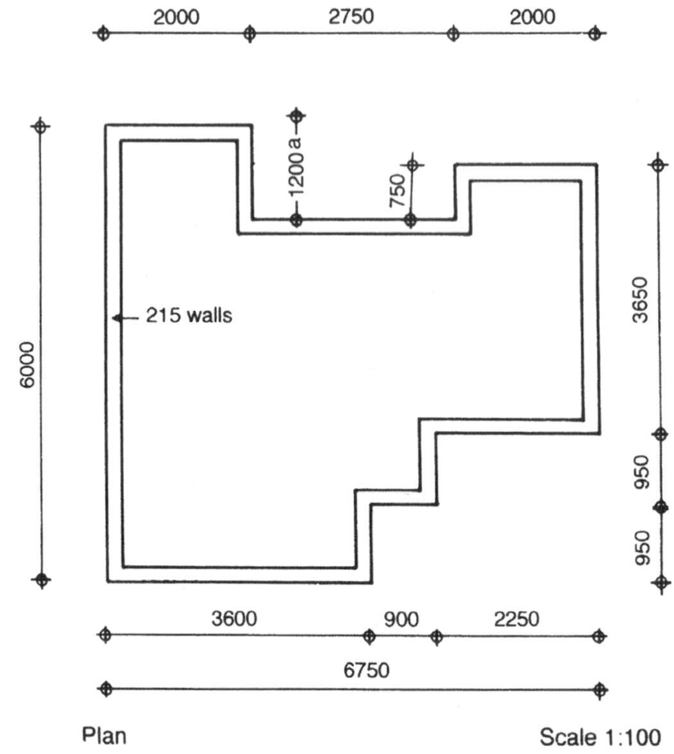

Fig. 8

Then, the calculation of the mean length is as follows:

	6 750
	6 000
	2/12 750
	25 500
2/750 =	1 500
	27 000
Less $4/2/\frac{1}{2}/215 =$	860
	26 140

If you have difficulty in appreciating why 2/0.750 is added, then colour the lengths of the wall that have already been measured carefully. This can make the extra length obvious. You could also think of this shape as having flexible joints, and 'push' the shape out to form a rectangle. Again, you will be left with two additional lengths that will not 'fit', these being the 2/0.750 for the 'indent' marked 'a' on Fig. 8.

The calculation of the mean girth is a most important one, as, having once been made, it is often used not only for several items in foundation measurement but also for brickwork and facings, with possibly copings, string courses and so on.

When calculating the girths of the individual components of a cavity wall, the same process should be followed. First the girth of the perimeter can be calculated; then, progressively, the mean girths of the outer skin, the cavity and the inner skin; and then, finally, the internal face of the inner skin (as in Example 3 in Chapter 9).

Chapter 7
General Principles for Taking-off

In this chapter, suggestions are given regarding procedure, general principles of measurement and other information that is necessary for the guidance of those responsible for coordinating the taking-off process, for the individual takers-off and for the study of the examples that follow.

It must not be forgotten that the methods of different individuals and the customs of different practices vary in the organisation and execution of this work. The fact that a particular suggestion is made here does not mean that it is universally adopted, nor does it preclude the use of an alternative.

A systematic approach is essential to ensure that the precise documents used when preparing a bill are known along with any revision numbers and the date of receipt. Recording who is dealing with which elements and what copies of drawings they are using helps with the management of the process, as various members of staff will refer to this record during the period of the project and not just the bill preparation stage. When variations are required at the post-contract stage or if a contractor challenges a quantity, the dimensions will need to be referred to.

Receipt of the drawings

As the drawings are received, they should be stamped with the date and entered on a list showing the drawing number, title, date of preparation, date received and number of copies. This list forms a useful reference and enables it to be seen when any revisions to previous drawings were received. A check should be made with the architect and engineer to ascertain what further drawings are under preparation that are likely to be received before the bill of quantities is completed. At the completion

Willis's Elements of Quantity Surveying, Twelfth Edition. Sandra Lee, William Trench and Andrew Willis.

Published 2014 by John Wiley & Sons, Ltd.

of the taking-off, each drawing used should be stamped 'used for quantities', which avoids confusion later on as to whether or not subsequent revisions were incorporated in the bill. When using CAD drawings, the version of the drawing used is just as important, and a schedule of drawings used for the bill is still required.

Preliminary study of drawings

Before any dimensions are entered or written at all, the taker-off should look over the drawings and study the general character of the building.

A check should be made to ensure that all floor plans and elevations are shown and that the positions of the sections are marked on all the plans. A more detailed inspection should be made to ensure that the windows, doors, rainwater pipes etc., shown on the plans are also shown on the elevations and vice versa. When drawings have been prepared by engineers, they should be compared with those prepared by the architect to ensure that there are no discrepancies in the layout or dimensions. If overall dimensions are shown, these should be checked against the total of room and wall measurements. If overall dimensions are missing, these should be calculated and marked on the plans, and all projections on external walls should be dimensioned. If this is done, the calculation of the mean girth and of the perimeter of the external walls is simplified. It is an advantage to dimension every room on the plans except where a series of rooms are obviously all of the same dimension in one direction. A little time spent in this preliminary figuring of the drawings obviates the possibility of inconsistency in dimensions. Where more than one taker-off is employed, the result of this work must be communicated to each of them, and their drawings marked accordingly. Figured dimensions must always be followed in preference to scaled, and any dimensions that can be calculated from those figured should be worked out. A larger scale drawing will usually override a smaller scale one, except when the smaller is figured and the larger is not. Practices vary when CAD drawings are used; however, enlarging drawings sufficiently to obtain an accurate measure is important.

Queries with the designers

Preliminary inspection of the drawings may also give rise to a number of queries to be raised with the architect or engineer, relating to missing information or discrepancies. Settling questions at an early stage saves interruption to the taking-off and consequently increases productivity. It is important that all queries are set down and answered in writing. Queries should be listed on the left-hand side of a sheet of paper, with the answers, when received, shown opposite on the right-hand side, together with the date and source of the answer (Fig. 9). The quantity surveyor, however, must accept some responsibility in making decisions, and many architects are quite ready to confirm such decisions. These query lists are then sent to the architect or the engineer for completion.

Project Nr	**Southtown School Queries**	Sheet Nr ....
Date		
Ref *Query*	*Reply*	*Date*
1. Finish to floor of entrance hall specified wood block, coloured as tile?		
2. Should dimension between piers on north wall be 5.08 not 5.03? (to fit the overall 56.85)		
3. Dpc not mentioned in the spec. notes?		
4. Should brick facing to concrete beams be tied back?		
Etc. etc.		

Fig. 9

If materials are shown for which full up-to-date information is not available in the office, it is advisable to ask manufacturers for literature and advice at an early stage. If the survey of the existing site does not show levels, or if those that are shown are insufficient, it is prudent to request that a grid of levels is taken over the site. These levels will prove invaluable when measuring earthworks and are essential for reference later if there is any question of re-measurement.

Initial site visit

Before actually starting on the measurement, it is advisable to make an initial visit to the site. If nothing else, such a visit allows one to get the feel of the job and to be able to visualise later the various site references when they are encountered on the drawings. A further visit or visits will be necessary when the measuring has progressed in order to pick up the spot items (see Chapter 19) and site clearance items (see Chapter 18). The initial visit will give an indication of what will be required in due course; in this respect, other items to be noted include such matters as boundaries generally (state and ownership of fences, walls, gates etc.), the existence or otherwise of overhead and underground services, means of access, adjoining buildings (historic, civic, defence etc.) and all matters that will be required for the subsequent drafting of the preliminaries bill (see Chapter 20). In addition, this visit is an opportunity to take the levels referred to here, and also to examine and note any trial holes that have been dug and are open for inspection. A photographic record of the site or specific parts of it may prove useful as a reminder of existing site conditions.

Where to start

If all drawings are completed and available, the taker-off will probably follow the order of sections given in this chapter or such other order as may be the custom of the office. This order will be seen to follow, more or less logically, the order of erection of the building, except that certain special work and services are dealt with at the end. It may, however, be that the design of, say, reinforced concrete foundations is not completed, and therefore measurement cannot begin as usual with substructure. In such a case a point, such as damp-proof course level, must be selected from which to measure the structure; the work below that level will be measured at a later stage when the necessary information is available. It may be that 1:100 scale drawings have been received, but 1:20 scale details are to follow. In such a case, internal finishes could probably be taken first, as the measurement of these is dependent generally upon the figured dimensions given on the 1:100 scale, which will probably not be altered. Moreover, these sections give the taker-off a good idea of the layout and general nature of the building. Specific forms of construction (e.g. a steel or concrete frame) will require a special sequence of measurement to be devised.

Organising the work

Where several takers-off are involved in the measurement of work and a deadline has to be met for the completion of the bill, it is important that the team leader organises the work carefully. To enable progress to be monitored, a schedule should be prepared showing the sections of work to be measured, with the name or initials of the taker-off responsible, together with the target and actual dates of commencement and completion of the work. Clear instructions must be given to each member of the group, carefully defining the extent of the work to be taken in each section. Proper arrangements should be made for the collection of queries for the architect or engineer, and these should be edited by the team leader before being passed to the other consultants. Proper supervision should be made of junior staff involved, and due allowance made for staff leave and their commitment to other work.

Sections of taking-off

The taking-off of dimensions is usually divided into sections under the main subdivisions of:

(0) Facilitating works (1) Substructure (2) Superstructure (3) Finishes (4) Fittings, furnishings and equipment (5) Services (6) Complete buildings and building units (7) Work to existing buildings (8) External works.

The measurement is often undertaken in elemental order, generally following the elements as laid out in NRM1:

0	Facilitating works	0.1	Toxic/hazardous/contaminated material removal
		0.2	Major demolition works
		0.3	Specialist groundworks
		0.4	Temporary diversion works
		0.5	Extraordinary site investigation works
1	Substructure	1.1	Substructure
2	Superstructure	2.1	Frame
		2.2	Upper floors
		2.3	Roof
		2.4	Stairs and ramps
		2.5	External walls
		2.6	Windows and external doors
		2.7	Internal walls and partitions
		2.8	Internal doors
3	Internal finishes	3.1	Wall finishes
		3.2	Floor finishes
		3.3	Ceiling finishes
4	Fittings, furnishings and equipment	4.1	Fittings, furnishings and equipment

5 Services	5.1	Sanitary installations
	5.2	Services equipment
	5.3	Disposal installations
	5.4	Water installations
	5.5	Heat source
	5.6	Space heating and air conditioning
	5.7	Ventilation
	5.8	Electrical installations
	5.9	Fuel installations
	5.10	Lift and conveyor installations
	5.11	Fire and lightning protection
	5.12	Communication, security and control systems
	5.13	Specialist installations
	5.14	Builder's work in connection with services
6 Complete buildings and building units	6.1	Prefabricated buildings and units
7 Work to existing buildings	7.1	Minor demolition works and alteration works
	7.2	Repairs to existing services
	7.3	Damp-proof courses/fungus and beetle eradication
	7.4	Facade retention
	7.5	Cleaning existing surfaces
	7.6	Renovation works
8 External works	8.1	Site preparation works
	8.2	Roads, paths, pavings and surfacings
	8.3	Soft landscaping, planting and irrigation systems
	8.4	Fencing, railings and walls
	8.5	External fixtures
	8.6	External drainage
	8.7	External services
	8.8	Minor building works and ancillary buildings

The headings listed here cannot be regarded as mandatory, but as they are RICS guidance, most surveyors would in fact follow these. In order to gain consistency and improve cost planning, NRM1 has identified what should be included in each element. It must be understood that the special requirements of any building may require additional or subdivided sections; the list given here must therefore be regarded as flexible.

Taking-off by work sections

Some surveyors make a practice of taking-off by work sections instead of by elements of the building as described in this chapter. As the final bill is usually arranged in work sections, this system can minimise the sorting or abstract, which, when measuring by sections of the building, is necessary to collect and classify the items into work sections. Such a system may present difficulties in some cases, and it would seem to increase the risk of duplicating items or of forgetting something. When measuring by sections of the building, the taker-off mentally erects the building step by step and is less likely to miss items. Sometimes, a combination of the two systems may be used to advantage.

Drawings

As mentioned in the 'Preliminary study of drawings' section, figured dimensions on drawings should be used in preference to scaling. Naturally, where there is a large discrepancy between the two, one should compare with other dimensions given, or, if this fails to produce a solution, consult the architect. Care must be taken to use the correct scale when taking measurements from drawings, particularly where a variety of scales has been used. As items are measured from drawings, it is often beneficial to loop through the applicable written notes and perhaps colour in the work on the drawing. When using CAD drawings, most software will provide coloured drawings to depict the items measured. Print-offs of these are useful when checking or editing the final bills. An examination of the drawings marked up in this way, at the end of measurement, will soon reveal any items not measured. Not all measurable items are shown on drawings, particularly if the drawings are incomplete or in any case for labours such as surface finishes to concrete, but the method should prevent any major items being overlooked.

The specification

If a specification is supplied, it should be read through cursorily first, not with the idea of mastering it in detail, but rather with a view to getting a general idea of its structure and content. A more detailed study should then be carried out of the sections relating to the element to be measured. For example, the excavation, concrete work and brickwork should be studied closely before beginning the taking-off for the substructure. It is useful when the taking-off is well advanced to go through the specification and run through in pencil all parts that have been dealt with, but not paragraphs that will form preambles to the bill or that contain descriptions that are not repeated in the dimensions. If this is done, it is unlikely that anything specified will be missed. Under the Joint Contracts Tribunal (JCT) form of contract, the specification is not a contract document; however, should the bill and specification both form part of the contract, they will have to be integrated. In such cases, the wording of the bill descriptions can be reduced by reference to the specification, particularly in a co-ordinated document. When the specification is a contract document, the preambles and the preliminaries in the bill can be similarly reduced in length.

More often than not, however, no formal specification is available at the outset. Brief specification notes may be supplied, and further notes must be made from verbal instructions given by the architect; these will form the nucleus of the specification, to be added to from time to time as queries are raised. These notes must be supplemented by the surveyor's own knowledge or by ideas of what is reasonably required. In making decisions on matters of specification, the surveyor should bear in mind that what is theoretically correct is not necessarily the most practical or economic solution.

Materials

The taker-off must have a thorough knowledge of the materials being used, and if a material is unknown should, where possible, make a point of seeing samples and studying the manufacturer's catalogue or leaflet, so that the limits and (perhaps optimistically stated) capabilities of the material are known. The handling of the material and study of any literature describing the materials and the way they are to be fixed in position will often assist in the measurement of the work or in the framing of a proper description.

Sequence of measurement

It is advisable when measuring to follow the same sequence in different parts of the work. For example, in collecting up the girth of the external walls of a building, it is a good idea to work clockwise, starting from, say, the top left-hand corner of the drawing. If this is done consistently, it will assist in reference later, when perhaps the length of a particular section of wall has to be extracted from a long collection. If a particular sequence of rooms has been adopted in measuring ceilings, the same sequence should be used for wall finish, skirtings, floors etc. In this way, if all the finishes of a particular room are to be traced, it will be known in which part of each group the relative dimensions are to be found.

Measurement of waste

It is normally stated as a general rule (though, like all rules, this has exceptions) that measurements of work are made to ascertain the net quantities as fixed or erected in the finished building. Wastage of material generally is allowed for by the contractor in the prices, though sometimes a measurement is made as a guide to the amount of waste, and in a few cases the gross quantity is measured. The exceptions to the general rule will be pointed out as they occur.

Overall measurements

It is usual in measuring to ignore in the first instance openings, recesses and other features that can be dealt with by adjustment later. Brickwork, for instance, is measured as if there were no openings at all, and deductions are made when the windows, doors or other openings are dealt with in the proper section. Plastering and similar finishes are measured in the same way. It simplifies the work to measure in this way. When, for instance, windows are being considered the sizes will be to hand, and openings will be measured to correspond. Moreover, it may happen that, say, internal plastering and windows are being dealt with by different takers-off, when it is obvious that the one who measures the windows is better able to make the deductions and adjustments. Most experienced takers-off decide that measuring overall with adjustment later is preferable to measuring net. This principle will also be found of value if, for example, a window should be forgotten, as the error would involve only the extra cost of the window over the wall and finishes, a much less serious matter than if nothing at all had been measured over the area of the window.

Use of schedules

Early preparation of information schedules of such items as finishes, windows and doors often exposes missing information and also provides a useful reference for the whole taking-off team. If like items are grouped together on the schedule, the taking-off process becomes more straightforward and the need to continually refer to the specification during the taking-off is avoided.

Use of scales

A warning should be given of the possibility of using the wrong scale in measuring. If possible, each side of the scale used should not have more than one variety of marking on each edge, but this is not always practicable. The scale most easily available for general use is a standard metric scale having 1:5, 1:50, 1:10, 1:100, 1:20, 1:200, 1:250 and 1:2500 markings. Some surveyors prefer to have a separate scale for each variety, or one marked on one face only; but when working on two different drawings (e.g. 1:100 and 1:20) at the same time – as is quite common – it is a great convenience to have both markings on the same scale. In any case, special care is necessary when measuring from different drawings to ensure that readings are taken from the correct markings of the scale. It may seem unnecessary to emphasise this, but mistakes on this account are not unknown.

Use of NRM2

The forms of contract agreed by the JCT provide that measurements shall be made in accordance with NRM2, and it is therefore of the utmost importance, where these forms of contract are used, that it should be followed. It should, nevertheless, be understood that, unless referred to in the contract, the SMM has no legal sanction and need not be adopted.

Decision on doubtful points

A thorough knowledge of NRM2 may still leave occasions when the taker-off must make decisions on the method of measurement or extent of descriptions for items not covered. When a rule of measurement is originated in such a way, it is often advisable to insert the method used as a preamble clause in the bill. When making such decisions, the taker-off should have one main consideration: what will best enable the estimator to understand (not merely guess) the work involved and enable it to be priced quickly and accurately?

Descriptions

The framing of descriptions so that they are both clear and concise is an art not easily acquired, but one which is of the utmost value. The contractor's estimator, always working at high pressure, will waste much time if faced with long-winded and rambling descriptions, or if having to decide what the surveyor intended to convey in a confused sentence. The surveyor, therefore, must always aim at clear expression, being careful in the choice of words and using the various technical terms in their proper sense. It is more common nowadays for standard description libraries to be used in which the descriptions are laid down for the taker-off. However, takers-off have to be very careful to avoid the pitfall of fitting an item to a standard description rather than ensuring that the description fits the item; if it does not, then a special or rogue must be compiled.

The requirements of NRM2 should be followed carefully when framing descriptions, but it must be remembered that additional information should be given where necessary to convey the exact nature of the work to the estimator. Whilst it may be convenient to follow the order of the tabulated rules, this does not prevent the use of traditional prose in the framing of descriptions. Certain items such as waste of materials, square cutting, fitting or fixing materials or goods in position, plant and other items are generally deemed to be included in descriptions. Where NRM2 calls for a dimensioned description to be given, apart from the description of the item, all dimensions should be given to enable the shape of the item to be identified.

The taker-off must also be careful to see that the same wording is used when referring to the same item in different parts of the dimensions, as inconsistency in the descriptions may indicate that some distinction must be intended by the different phraseology. For instance, if the taker-off, describing plaster, writes '2 coat plaster on block walls', and then after several such items writes later '2 coat plaster on partitions' they may end up as different items in the bill, although this was not the intention of the taker-off. Therefore, when the same item appears in different places, it should be written in exactly the same form, or after the first time abbreviated with the letters 'a.b.' (for 'as before'; discussed further in the 'Abbreviations' section) to indicate that exactly the same meaning as before is intended. It is important to confine the description to what is actually to appear in the bill, and not to add to it particulars of location or other notes that are merely for reference and not intended to go any further.

The descriptions written on the dimension sheets should be in the form that they are intended to appear in the bill, with such parts that would normally be covered by a preamble (see Chapter 21) being omitted. Notes to assist in writing the preambles, particularly for the less common items, can be entered on the dimension sheet but should be kept well clear of dimensions or descriptions.

In some cases, there will be added to the description of a superficial or linear item a note of the number included in the item so that the estimator can judge the average size of each, for example:

3/2/	3.50 1.60		Formwork, fine finished formwork to sides of attached columns, regular square shape. (6 no.)

Some of the requirements of NRM2 for descriptions may be covered by general clauses or preambles to each work section. For example, NRM2 Section 17 on roof coverings says that the lap in joints of flashings shall be given. A preamble clause saying that all flashings are to be lapped 100 mm at joints would be sufficient and would probably save repetition.

Abbreviations

Under the traditional method of taking-off, abbreviations can be extensively used to shorten descriptions and individual words. In the examples in this book, abbreviations have been kept to a minimum.

A special note might perhaps be made here of the abbreviation 'a.b.' for 'as before'. Where this is used and might refer to more than one item, it always refers to the last such item. For instance, a description '44 mm door a.b.' would refer to the last type of 44 mm door measured, if there had been several different varieties. If, however, there is any doubt, it is best to add to the brief description sufficient for it to be identified or to say '44 mm door a.b. col. 146', the reference to the column number being a definite guide.

Extra over

Some items are measured as *extra over* others, that is, they are not to be priced at the full value of all their labour and materials, as these have to a certain extent already been measured. For example, labours on structural steelwork such as cranks to beams are measured as extra over.

This means that the beam is measured its full length over the cranked bend and the estimator, when pricing the item, assesses the extra cost for forming the crank. The measurement of an item as extra over something already measured as extra over should be avoided.

Dimensioned diagrams

There is a requirement in NRM2 for a certain amount of drawn information to be made available. Most of this drawn information will be available from the drawings used for the taking-off, and so the requirement only means ensuring that the information required is included and that the requisite number of copies of the various drawings accompany the tender documents. However, in certain cases NRM2 suggests dimensioned diagrams or sketches to elaborate a description. These diagrams or sketches can either be produced separately for inclusion in the text of the bill of quantities or drawn by the taker-off, with the dimensions to be processed as a separate operation.

These diagrams will be either extracts from the drawings or drawings specially prepared for incorporation, with the written description on the facing page. Alternatively, the diagrams are collected and printed on one or more sheets at the end of the bill, each diagram being given a number and being referred to by that number in the body of the bill. Diagrams should be marked 'plan', 'section' and so on and be drawn to scale, the scale being indicated or the dimensions figured. For example, NRM2 Section 11.17.2 calls for a dimensioned diagram to clearly describe a complex formwork shape of a beam.

Apart from what is discussed here, sketches will be made in the dimensions or on separate sheets to work out points of construction not detailed or to supplement the architect's details. If possible, these should be made on the dimension sheets where the particular work is measured, on a spare space on the drawings, or if made on separate sheets they should be carefully preserved to show at a later stage what the taker-off had assumed when measuring. Any details of importance should be confirmed with the architect.

Prime cost items and provisional sums

For various reasons, it is not always possible when quantities are being prepared to define finally everything necessary for the completion of the building. For instance, it may be necessary for the architect to select certain articles, such as sanitary appliances, ironmongery and so on in consultation with the client, and the details of these may very well not have been considered at the early stage when tenders were being obtained. It is not unusual, therefore, to put in the bill *prime cost* (PC) sums for these items, which the estimator will include in the tender for goods to be obtained from a supplier, but which are subject to adjustment against the actual cost of the articles selected. The contractor is given the opportunity in the tender to add for profit to each of these items.

Provisional sums are sums included for risk items and work to be done by the general contractor for which there is insufficient information for them to be adequately described in the bill of quantities. How PC and provisional sums are incorporated in a bill of quantities is dealt with in Chapter 20.

Approximate quantities

Work to be carried out by the general contractor, which may be uncertain in extent, can also be provided for by means of approximate quantities (i.e. by measuring work in the normal way but keeping it separate in the bill and marking it 'approximate'). For instance, the foundations of a building, where the nature of the soil is uncertain, may be measured as shown on the drawings, and additional excavation, brickwork and so on measured separately and marked 'approximate' to cover any extra depth to which it may be necessary to take the foundations. It is thereby made clear that adjustment of these quantities on completion of the work is anticipated. Alternatively, if there is a considerable amount of such work, it may be contained in a bill of approximate quantities. As noted in the 'Prime cost items' section, if the work cannot be described adequately, it is to be included as a provisional sum.

Summary

(1) Prepare a drawing register, and log any revisions received during bill preparation.
(2) Check the plans, sections and elevations for any clashes of information or missing details.
(3) If measuring manually, put headings onto the dimension paper and number sheets. If using a computer system, structure the sections to be measured according to the structure of the final bill requirements.

(4) Measure work from drawings in a set, logical pattern. One method frequently adopted is to start in the top left-hand corner of the drawing and work clockwise.
(5) Write clearly and legibly, and space your work out.
(6) Start by producing a query sheet and then a taking-off list for each section.
(7) Measure items in the same sequence as your taking-off list, using figured dimensions in preference to scaling from the drawings.
(8) Waste calculations should appear before descriptions.
(9) Give locational notes against your dimensions so that work can be followed easily.
(10) Use schedules where possible to save repetition.
(11) Use 'to take notes' when you do not have sufficient information to measure an item, but make sure to check these prior to final bill production.
(12) When your measure is complete, go back over the drawings and line through all of the items measured. Then repeat this process with your take-off list to ensure that you have not missed an item.
(13) Check balances to ensure, for example, that all excavated material has been disposed of or used as filling where appropriate.
(14) Stamp drawings as used for bill preparation.

Chapter 8
Substructures

Particulars of the site

Before beginning the measurement of substructures, the drawings must be examined to ascertain whether the existing ground levels are shown in sufficient detail for calculating average depths of excavation. If the levels are not shown or are insufficient, then it is necessary to take a grid of the levels over the site. Irrespective of taking levels, the surveyor should always visit the site to ascertain the nature and location of existing buildings, and details for preliminaries items and for the measurement of excavation work. Among items to be noted for the latter are vegetation to be cleared, the existence of topsoil or turf to be preserved, pavings or existing structures in the ground to be broken up and, if trial holes have been dug, the nature of the ground and the groundwater level. Where the proposed work consists mainly of alterations, an early visit to the site will be essential, and most of the taking-off may even have to be carried out there. The measurement of alterations or *spot items* and the methods of dealing with this class of work are dealt with in Chapter 19.

Bulking

When measuring excavation, disposal and filling, the dimensions are taken as in the ground, trenches being measured along their centre line and then multiplied by their width and depth as shown on the drawing. Soil increases in bulk when it is excavated, but no account is taken for this in the bill of quantities, with the estimator having to make the due allowance.

Willis's Elements of Quantity Surveying, Twelfth Edition. Sandra Lee, William Trench and Andrew Willis.

Published 2014 by John Wiley & Sons, Ltd.

Removing topsoil

Where new buildings are to be erected on natural ground, it is necessary to measure a separate superficial item for the stripping of the vegetable soil or topsoil where it is to be preserved. This is measured over the area of the whole building including the projection of concrete foundations beyond external walls. A separate cubic item has to be taken for the disposal of the topsoil, giving the location. Any further excavation for trenches, basement and so on would then be measured from the underside of such topsoil excavation. If there are existing paths, paving, and the like over portions of the area to be stripped, an item must be taken for breaking out existing hard pavings, as a superficial item stating the thickness and the material and measuring the removal as a separate item. Breaking out may be taken as extra over the excavation.

Where the site is covered by existing buildings, the pulling down is dealt with as described in Chapter 19, no item for stripping topsoil being necessary. Demolition is usually taken to the existing ground level; breaking out below ground level is measured as a cubic item which may be taken as extra over the excavation.

Where the site is covered with good turf which is worth preserving, an item should be taken for lifting it (NRM2 5.1.1) and a separate item for relaying any turf to be reused (NRM2 37.4.1).

Bulk excavation

Where a site is sloping, it is often more economical to set the ground-floor level so that one end of the site must be excavated into, that is, the underside of part of the hardcore bed will be below the level of the ground after the topsoil is stripped. Where this is the case, a cubic measurement is made of the excavating to reduce levels necessary from the underside of the topsoil excavation already measured to the underside of the hardcore bed. The depth for this item must be averaged, and it will generally be found that excavation is only necessary over part of the site, the level of the remainder being made up with hardcore or other filling. In Fig. 10, it can be seen that the ground level must be reduced to a formation level of 44.00 (300 mm below the top of the floor slab). The contour of 44.00, known as the *cut and fill line*, is plotted on the plan as accurately as possible from the levels given; the area on the right-hand side of this contour is measured for excavation, and that on the left-hand side for filling.

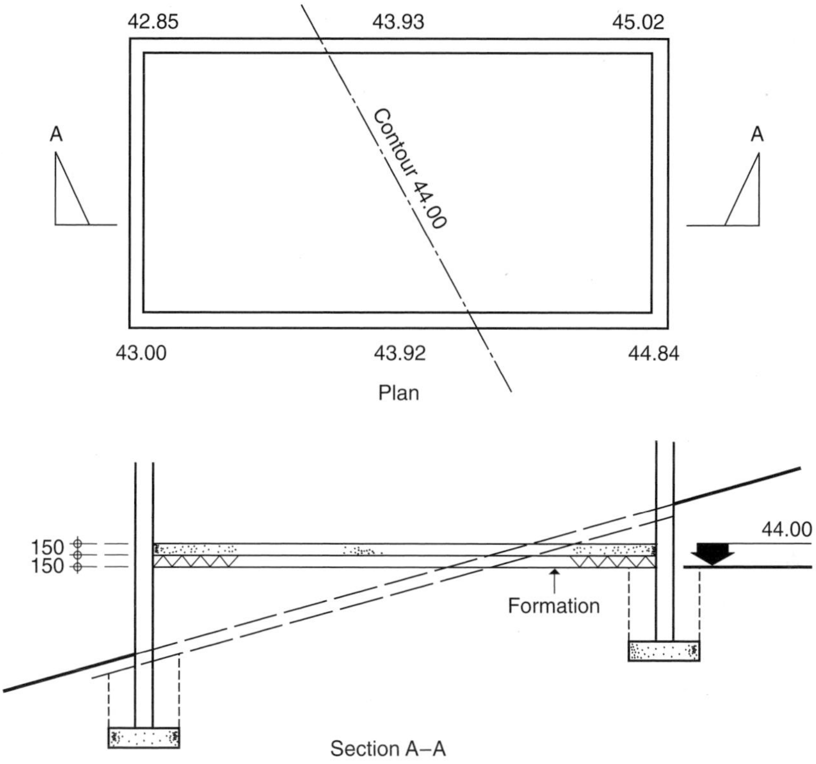

Fig. 10

Additional contours may have to be plotted to classify the excavation according to the maximum depths required by NRM2. It may, however, be more sensible to find an average depth for the whole excavation for measurement purposes and classify the depth in the description as the maximum on site. If this method is used, a statement should be made in the bill to this effect.

The bulk excavation to reduce levels must be measured before any foundation excavation, as it brings the surface to the 'reduced or formation level' from which the foundation excavation is measured, and, like the stripping of topsoil, it must be measured to the extreme projection of concrete foundations. A separate cubic item for disposal of excavated material must also be taken. It will often be necessary, where the floor level of the building is below the ground outside, to slope off the excavation away from the building, and possibly to have a space around the building excavated to below the floor level. In such circumstances, the additional excavation would be measured with the external works section.

Excavation for paths

Stripping of topsoil and bulk excavation to reduce levels may also be required for formation of paths, paved areas, and so on. External work of this nature is best measured all together after the building has been dealt with, as it usually forms a separate section in the bill. When paths abut the building (thus overlapping the projection of foundations), the whole excavation necessary for the erection of the building should be measured with the building, the extra width necessary for the paths only being measured with the paths.

Levels

Before foundations are measured, three sets of levels must be known:

(1) Underside of concrete foundation
(2) Existing ground level
(3) Floor level.

(1) and (2) are necessary to measure trench foundation excavation, and (1) and (3) are necessary to calculate correct heights of brickwork or other walling. The natural ground level, as has been pointed out, will probably vary and have to be averaged either for the whole building or for sections of it; if the floor level and the underside of the foundation are constant, the measurement of trench foundation excavation is fairly simple. However, both the underside of the foundation and the floor level may vary in different parts of the building, there being steps at each break in level, and sometimes the measurement of foundations thus becomes very complicated. It will be found useful to mark on the plan the existing ground levels at the corners of the building – if necessary, these can be interpolated from given levels – and in the same way the levels of the underside of the foundation can be marked on the foundation plan (if any). If stepped foundations are required, it will prove helpful if the foundation plan is hatched with distinctive colours to represent the varying depths of the underside of the foundations. If no foundation plan is supplied, the outlines of foundations can be superimposed on the plan of the lowest floor.

Foundation excavation

The measurement of foundation excavation, in particular trench foundation work, will normally divide itself into two sections:

(1) External walls
(2) Internal walls.

The former is dealt with first. In the simplest type of building, a calculation is made of the mean length of the trench, as described in Chapter 6. This mean length of the trench will also be the mean length of the concrete foundations. The width of trench will be the width of the concrete foundation as shown on the sections or foundation plan. The depth of trench will, if the underside of concrete foundation is at one level, be the difference between that level and the average level of the ground, after making allowance for stripping of the topsoil or bulk excavation to reduce levels already measured. Where the underside of concrete is at different levels, theoretically the excavation for each section of foundation between steps should be measured separately, the lengths when measured being collected and checked with the total length ascertained. However, it may be found in practice that, the steppings to bottom of the trench being small and the ground normally falling in the same direction, an average depth can be determined for larger sections of the building, if not for the whole. The depth of trench foundation excavation is given in 2 m stages. Alternatively, as mentioned in this chapter, in bulk excavation the maximum depth may be given and a statement made in the bill giving the method used. The main measurement of excavation for the external walls having been set down, excavation for any projections on this foundation can be measured.

It has been assumed here that the external walls have strip foundation trenches of a uniform width all round. Trench excavation is measured as a cubic item, and a separate cubic item of disposal must be taken.

Internal walls must be collected up in groups according to their average depths. Allowance should be made in the length for overlap where an internal wall abuts against an external wall, by deducting from the length of the internal wall the projection of concrete foundation to the external wall at this point. A similar allowance should be made where internal walls intersect. The necessity for this is best shown diagrammatically (Fig. 11).

It will be seen that if the foundation for the internal wall were measured the same length as the wall (i.e. as dotted in Fig. 11) the area marked with a cross would be measured twice for the excavation and concrete. The amount involved being comparatively small, some surveyors may ignore the deduction, but to do so when there are many intersections is to take an unnecessary and definite over measurement without any justifiable reason.

Students often find it difficult to decide whether the maximum depths for trench excavation and the like should be calculated from the original

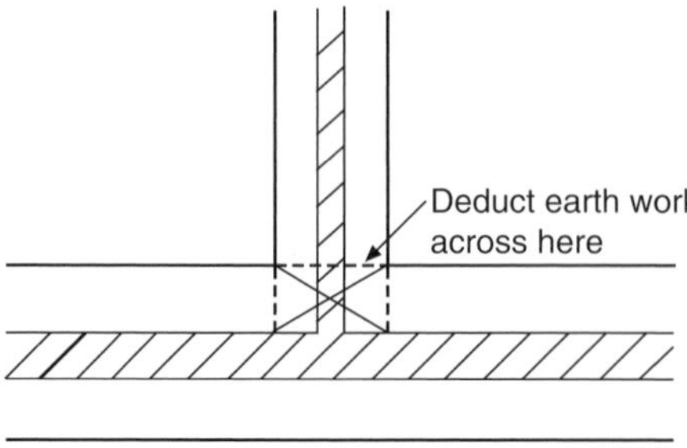

Fig. 11

ground level or from the level of the ground after topsoil has been stripped, where this is measured separately. It is usual to measure from the latter level as this is the commencing level of the actual excavation being measured. NRM2 requires that the commencing level of excavation must be stated where this is not the existing ground level, the depth classification of excavation being given from the commencing level.

Earthwork support

Earthwork support is not measurable under NRM2, except where specifically designed and required by the engineer.

Disposal of excavated material

It must be ascertained how the excavated material is to be disposed of. It is naturally cheaper if the material can be disposed of on the site, but there is often no room for it and it must then be removed from site. The special circumstances of each case must therefore be considered, and the disposal fully described accordingly. Care must be taken to see that every item of excavation in the dimensions has an appropriate item of disposal measured. Where part is to be retained and part removed from site or otherwise disposed of, it will be found simplest to measure an additional item in the first instance as filling with material arising from the excavations equal to the volume of excavation as marked by the cross on Fig. 12. When concrete and brickwork are measured later, an adjustment can be made of the volume occupied by these as:

Deduct Filling to excavations final thickness exceeding 500 mm

&

Add Disposal of excavated material off-site.

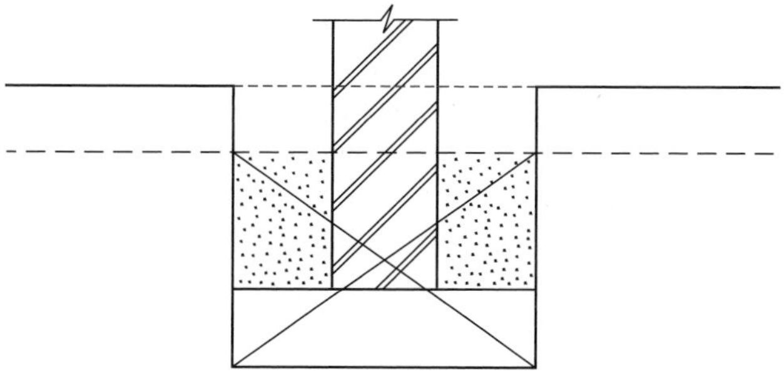

Fig. 12

If filling is deducted for the volume occupied by concrete and brickwork (shown hatched in Fig. 12), whatever is remaining of the original measurement will be the volume of space to be filled. It is simpler to calculate the volume of the brickwork than to obtain the volume of the spaces on each side. This is one example of the advantage of the overall system of measurement. Since all is measured in the first instance as filling, if adjustment for removal is forgotten, the error is less serious. Care is necessary in making the adjustment for the volume occupied by the wall to ensure that the height is not taken above the level to which the filling was measured in the first instance. In the case of basement excavation, it is probably more convenient to measure all for removal, and adjust subsequently for filling round the outside.

If the filling round the walls is to be hardcore, then this would be taken in the first instance, instead of filling with material arising from the excavations, together with a removal item for the excavated material. If only the inside of the trench is filled with hardcore, then an adjustment will have to made for the filling to the outside, as demonstrated in Example 1.

Working space

Working space is not measurable under NRM2.

Concrete foundations

The length measured for excavation of strip foundation trenches will usually be found to serve for the centre line of the concrete measure. The width will be the full width of trench, and the thickness as shown on the drawings. Where the concrete foundation is not at a uniform level throughout, it will be found easiest to measure it as if it were, the additional concrete being added afterwards for the laps together with temporary support known as *formwork* to the face of steps. Again, the advantage of overall measurement is apparent; if each section were measured piecemeal, there would be more danger of error through a section being missed. Concrete poured on or against earth or unblinded hardcore has to be so described.

Concrete foundations are sometimes reinforced by either steel fabric or bars. In such cases, it is important to remember that a finer aggregate is necessary than in ordinary mass concrete, and the concrete will therefore be of a different composition but may otherwise be measured in the same way except that it is described as reinforced. A separate linear item is taken for the reinforcing bars, which is weighted up in the bill. Fabric reinforcement is measured as a superficial item, stating the width if in a one-width strip.

When foundations are reinforced, weak concrete blinding 50 or 75 mm thick is usually required under the foundation. Such blinding is sometimes not shown on the drawings, in which case enquiry should be made of the architect or engineer as to whether it is wanted.

Concrete ground slab

Hardcore or other filling to make up levels under concrete ground slabs is measured as a cubic item stating the average thickness as being either less than or greater than 500 mm. Blinding the surface of the filling is measured superficially, stating the material to be used. Damp-proof membranes are also measured superficially as exceeding 500 mm in respect of the area in contact with the base, with no deductions being made for voids less than 1.00 m^2.

Concrete ground floor slabs are measured as cubic items stating the thickness as not exceeding 300 mm or as exceeding 300 mm; the thickness excludes projections or recesses. When reinforced, this must be stated; slabs poured on or against earth or unblinded hardcore must be measured separately. Non-mechanical tamping of concrete surfaces is deemed included; however, other surface treatment such as power floating is measured separately.

Formwork is measured to the edges of beds and given linearly, stating the height as not exceeding 500 mm or as exceeding 500 mm. Fabric reinforcement to slabs is measured in square metres.

Brickwork and blockwork in foundations

This work is often kept under a separate heading of *substructure.*

Where the concrete foundation is stepped, each length of wall from step to step will have to be measured separately up to damp-proof course level. As this is piecemeal measurement the individual lengths taken should be totalled and compared with the mean girth of the entire wall. Alternatively, the total length may be measured by the minimum height, the extra heights being added for each section. When measuring internal walls, it should be noted that their length will not be equal to that taken for their concrete foundations as the length of these is adjusted at intersections as mentioned in this chapter.

Brickwork below damp-proof course level is commonly specified to be in cement mortar, as opposed to gauged mortar for brickwork above, and this must be made clear in the descriptions. The wall between ground level and the damp-proof course is often constructed of facing bricks, which may be continued for one or two courses below ground level to allow for irregularities in the surface of the ground. The measurement of walling in substructure follows the rules for the measurement of general walling, covered in Chapter 9.

Damp-proof courses

These are measured as superficial items and classified by width as exceeding or not exceeding 300 mm. Horizontal, raking, vertical and stepped damp-proof courses are each kept separately. Pointing to the exposed edges is deemed to be included, but the thickness of the material, the number of layers and the nature of the bedding material have to be stated. In the measurement, no allowance is made for laps. Asphalt damp proofing and tanking are measured using the area in contact with the base as a superficial item, with the widths exceeding or not exceeding 500 mm. The thickness, number of coats and nature of the base and any surface treatment have to be included in the description together with the pitch. Internal angle fillets ends and angles fair edges, rounded edges and arrises (or external angles) are deemed to be included. Raking out joints of brickwork for a key is deemed to be included with the brickwork. No deduction is made for voids in asphalting not exceeding 1 m^2.

Example 1

Wall Foundations

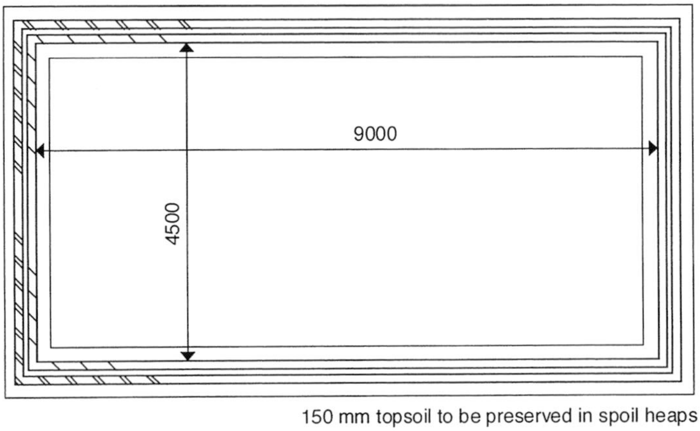

PLAN Scale 1:100

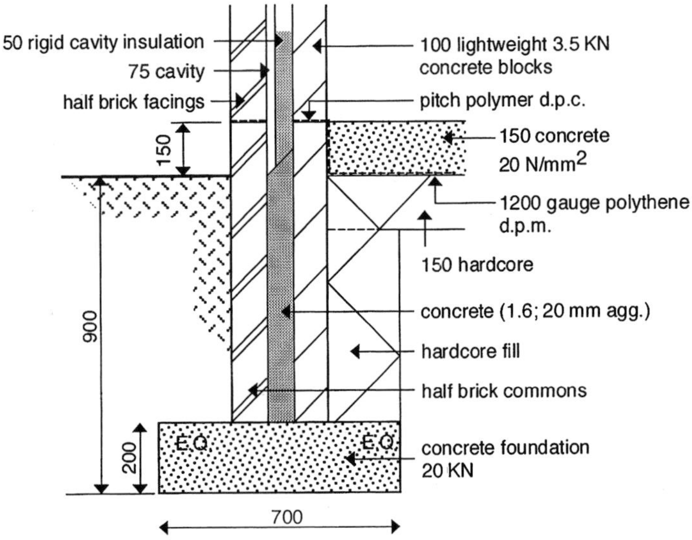

SECTION Scale 1:20

Fig. 13

Example 1

Taking-off list	**NRM2 reference**
Topsoil excavation	5.5.2
Retaining topsoil	5.10.1
Foundation excavation	5.6.2
Disposal of soil	5.9.2
Surface treatment	Not measurable
Earthwork support	Not measurable
Concrete foundation	11.2.1
Cavity wall: brick	14.1.1
block	
form cavity	14.14.1
Cavity fill	11.5.1
Adjust soil disposal	5.9.2
Hardcore fill	5.12.3
Topsoil backfill	5.11.2
Damp-proof course	14.16.1.3
Adjust for facings	14.1.1
Concrete bed	11.5.2
Hardcore bed	5.12.3
Surface treatment to hardcore	Not measurable
Damp-proof membrane	5.16.2
Membrane upturn	5.16.1
Concrete treatment	11.8.1
Ground treatment	Not measurable

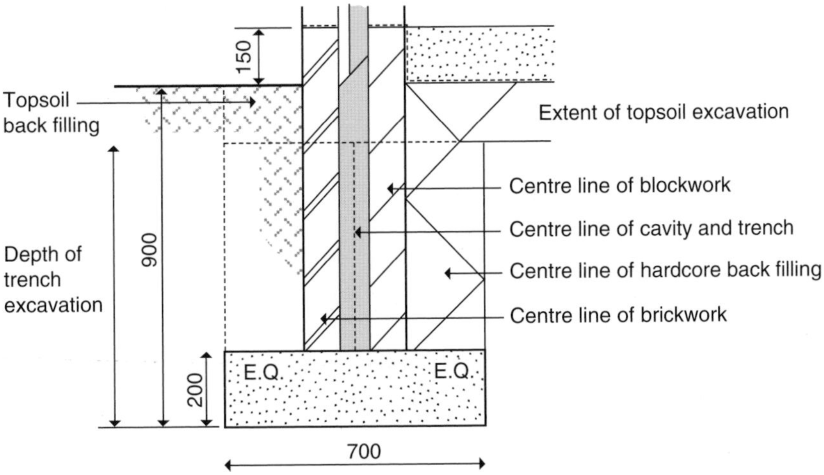

Fig. 14

				Substructure 1
			Half brick wall 102.5 Cavity 75 Block skin <u>100</u> <u>277.5</u>	Calculation of wall thickness
			Trench width 700 Less wall <u>277.5</u> 2)422.5 Foundation spread 211.25	
			<u>W</u> <u>B</u> 9.000 4.500 0.555 2/277.5 0.555 <u>0.422</u> 2/211 <u>0.422</u> <u>9.977</u> <u>5.477</u>	Calculation of overall dimensions to outer edge of concrete foundation
	9.98 <u>5.48</u>		Site preparation, removal of topsoil average 150 mm thick	5.5.2 Only measured if required to be retained
			&	
			Retaining excavated material on site, topsoil, in temporary spoil heap average 100 m from excavation <u>x 0.15 = m³</u>	5.10.1.2 Location required if not at discretion of contractor

				Substructure 2
			Depth of trench	
			0.900	Label each one clearly.
			less topsoil 0.150	
			0.750	
			Mean girths	
			9 000	
			4 500	
			2/13 500 27.000	
			4/2/½/277.5 1.110	
			Centre line of trench 28.110	
	28.11		Excavation commencing 150 mm below original ground level, foundation excavation maximum depth n.e. 2.00 m	5.6.2
	0.70		&	
	0.75		Disposal of excavated material off-site	5.9.2 This will be adjusted later for backfilling.

				Substructure 3
	28.11 0.70 0.20		In-situ concrete (20 N/mm^2) horizontal work, less than 300 mm thick in structures, foundation, poured against face of earth	11.2.1.1 Topsoil depth 150 mm 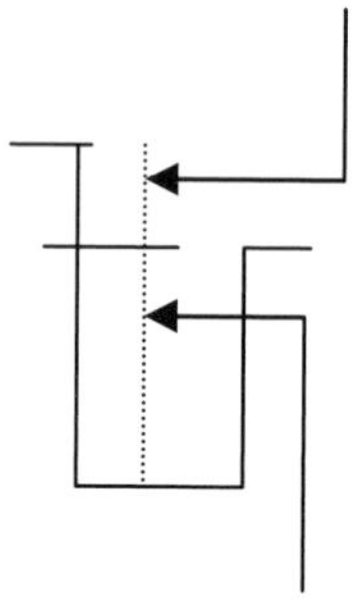 Trench depth 0.75 m
			<u>Height of brickwork</u> Depth of foundations 0.900 To damp-proof course (d.p.c.) <u>0.150</u> 1.050 less concrete depth <u>0.200</u> <u>0.850</u>	
			<u>Girths</u> ½/102.5 51.25 27.000 cavity 75 blocks <u>100</u> 4/2/ 226.25 <u>1.810</u> c.l. brickwork <u>28.810</u>	These girths are taken from the internal girth as calculated before and using the 4/2/ distance moved principle.
			27.000 4/2/½/100 <u>0.400</u> c.l. blocks <u>27.400</u>	It is assumed all walls are vertical unless otherwise stated.

				Substructure 4
	28.81 0.85		Common brickwork, Walls, half brick in stretcher bond, skins of hollow walls in cement mortar (1:4)	14.1.1 Outer skin brickwork up to dpc
	27.40 0.85		Dense Concrete Blocks, Walls 100 mm thick in skins of hollow walls, cement mortar (1:4)	14.2.1 Inner skin blocks
	28.11 0.85		Forming cavities in hollow wall, 75 mm wide, including stainless steel twisted wall ties, 5 per square metre.	14.14.1 State width in description
			0.900 less foundation 0.200 0.700 plus top splay 0.050 0.750	Depth of cavity filling
	28.11 0.08 0.75		In situ concrete (1:6 – 20 mm aggregate) vertical work, ≤300 mm thick, filling to hollow wall.	11.5.1.1

				Substructure 5
			Trench depth 0.750 Less concrete 0.200 0.550	Depth of backfilling
			half spread ½/211.25 106 27.000 wall 278 4/2/ 384 3.072 30.072	
	30.07 0.21 0.55		Deduct Disposal excavated material off site & Add Filling to excavations exceeding 0.5 m thick with excavated material	5.9.2 This is the adjustment for earth filling to the outside of trench 5.11.2
			Filling 27.000 Less 4/2/½/211.25 0.845 26.155	
	26.16 0.21 0.55		Imported filling, hardcore, backfilling around foundations, exceeding 0.5 m thick	5.12.3 Inside filling

				Substructure 6
	30.07 0.21 0.15		Filling to excavations n.e. 0.5 m thick with retained topsoil from site spoil heap n.e. 10 m distance, lightly compact.	11.1.1 Topsoil replacement outside buildings
	28.81 0.10 27.40 0.10		Damp proof courses n.e. 300 mm wide, horizontal, pitch polymer, bedded in gauged mortar (1:1:6)	16.1.3
			Facings Three courses bwk 3/0.075 0.225	Taken one course below ground level and up to dpc
	28.81 0.23		Deduct Walls in commons as before & Add Facing brickwork, Walls, half brick thick, skins of hollow walls, stretcher bond in gauged mortar (1:1:6) including pointing with a weathered struck joint as the work proceeds.	14.1.1 The type of facing brick should be stated

				Substructure 7
			Ground floor slab	
	9.00 4.50		In situ concrete (C20 N/mm^2 -20 mm aggregate) horizontal work ≤ 300 mm thick	11.5.2
			x 0.15 = m3	
			&	
			Imported filling, hardcore, to make up levels n.e. 0.5 m thick, imported hardcore compacted inlayers 150 mm thick	5.12.2
			x 0.15 = m3	
			&	
			Surface treatment to concrete, trowelling	11.8.1
			&	
			Damp proof membrane exceeding 500 mm wide, horizontal, 1200 gauge polythene laid on blinded hardcore to receive concrete	5.16.2.1
	27.00		Damp proof membrane width n.e. 500 mm of 1200 gauge polythene sheeting vertical at abutment	5.11.1 Vertical turn up of damp proof membrane. Inside girth of blockwork,

Example 2

Basement Foundations

(4 m square internally on plan)

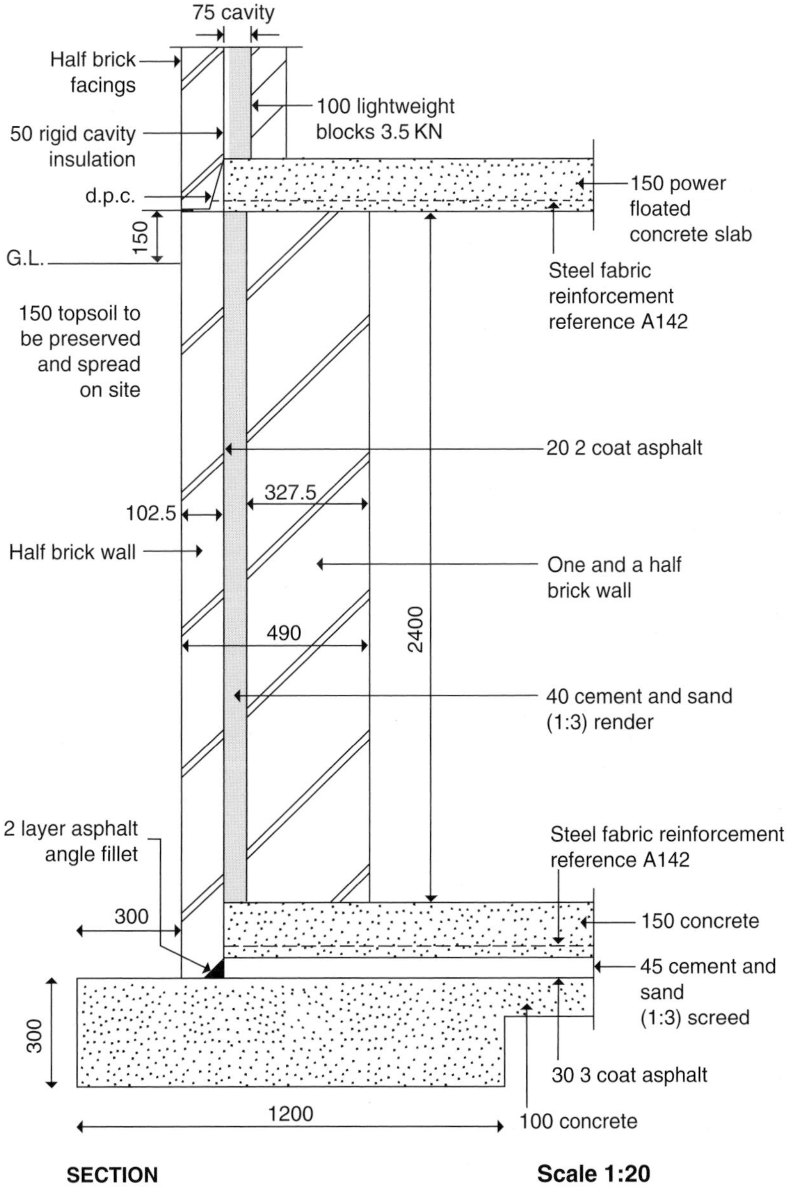

Fig. 15

Example 2

Taking-off list	NRM2 reference
Topsoil excavation	5.5.2
Retaining Topsoil	5.10.1
Surface treatment: bottom of excavation	Not measurable
Bulk excavation	5.6.2
Disposal of soil	5.9.2
Earthwork support: basement	Not measurable
Foundation excavation	5.6.2
Disposal of soil	5.9.2
Earthwork support: basement trench	Not measurable
Working space: basement	Not measurable
Trench foundation: concrete	11.1.1
Concrete bed	11.1.1
Surface treatment: concrete	11.8.1
Asphalt tanking: horizontal	19.1.1
Floor screed	28.1.2
Concrete bed	11.2.1
Mesh reinforcement	11.37.2
Formwork to edge of bed	11.14.1
Inner structural wall	14.1.1
Wall render	28.7.2
Asphalt tanking: vertical	19.1.3
wall damp-proof course (d.p.c.)	19.2.1
angle fillet	Not measurable
edges	Not measurable
Half brick protective wall	14.1.1.4
Concrete suspended slab	11.2.1.2
Mesh reinforcement	11.37.2
Surface treatment: concrete	11.8.1
Formwork to slab: soffits	11.15.1.1
edge	11.14.1
Cavity tray d.p.c.	14.17.2.3
Adjust soil disposal: backfill	11.2.1
remove from site	9.2.1
topsoil	11.1.1
Adjustment for facing bricks	14.1.1

			<u>Girths</u>	**Basement 1**
			Wall 490 4000	**4 m square internally on plan**
			Spread <u>300</u>	
			2/790 <u>1580</u>	As you become more familiar with calculating girths you can work out all the required girths at the start of your take-off and label for later use.
			4/5580	
			Outer girth 22 320	
			- 4/2/½/300 <u>1 200</u>	
			Backfill 21 120	
			- <u>1 200</u>	Calculations of overall dimensions to outer edge of concrete.
			Outer face 19 920	
			-4/2/½/102½ <u>410</u>	
			19 510	
			- <u>410</u>	
			19 100	
			-4/2/20 <u>160</u>	
			Asphalt 18 940	
			-4/2/40 <u>320</u>	
			Render 18 620	
			-4/2/½/327½ <u>1 310</u>	
			1 ½ brick 17 310	
			- <u>1 310</u>	
			Inner girth <u>16 000</u>	
			Outer girth 22 320	
			-4/2/½/1200 <u>4 800</u>	
			trench girth <u>17 520</u>	
			<u>Depths</u>	
			2 400	
			Concrete 150	
			Screed 45	
			Asphalt 30	
			Concrete <u>100</u> <u>325</u>	
			2 725	
			less dpc - ground <u>150</u>	
			2 575	
			less topsoil <u>150</u>	
			<u>2 425</u>	

				Basement 2
				Basement excavated to underside of concrete bed
	5.58 5.58		Site preparation, removal of topsoil average 150 mm thick	5.5.2 Only measured if required to be retained
			&	
			Retaining excavated material on site, topsoil, in temporary spoil heap average 100 m from excavation x 0.15 = m^3	5.10.1.2 Location required if not at discretion of contractor
			&	Surface treatment deemed included
			Bulk excavation commencing 150 mm below original ground level, basement excavation maximum depth n.e. 4.00 m x 2.43 = m^3	5.6.1.2
			&	
			Disposal of excavated material off-site x 2.43 = m^3	5.9.2
	17.52 1.20 0.20		Excavation commencing 150 mm below original ground level, foundation excavation maximum depth n.e. 2.00 m	5.6.2
			&	
			Disposal of excavated material off-site	5.9.2

				Basement 3
	17.52 1.20 0.20		In-situ concrete (20 N/mm^2) horizontal work, less than 300 mm thick in structures, foundation, poured against face of earth	11.2.1.1
	5.58 5.58 0.10		In-situ concrete (20 N/mm^2) horizontal work, less than 300 mm thick in structures, bed, poured against face of earth	11.2.1.1
			Tanking 4 000 Wall 2/327 ½ 655 Render 2/40 80 4 375 Asphalt 2/20 40 4 775	
	4.78 4.78		Trowel surface of concrete to receive asphalt	11.8.1
			&	
			Mastic asphalt tanking/damp-proof membranes with limestone aggregate to BS 6925 finished with a wood float using fine sand as abrasive; subsequently covered	
			Tanking and damp proofing width exceeding 500 mm, 30 mm thick in three coats laid horizontal on concrete	19.1.1

				Basement 4
			Outer wall 2 400 Concrete 150 Screed 45 Asphalt 30 2 625 Less asphalt damp-proof course (d.p.c.) 20 2 605	
	19.51 2.61		Common brickwork, walls, half brick in stretcher bond, built against face of asphalt, in cement mortar (1:4)	14.1.1.4
			Inner wall	
	17.31 2.40		Common brickwork, walls, one and a half brick thick in English bond, fair and finished flush pointed one side as the work proceeds, in cement mortar (1:4)	14.1.1
	18.62 2.40		Cement and sand (1:3) render to brick walls width exceeding 300 mm, 40 mm thick in three coats	28.7.2
	18.94 2.61		Asphalt tanking as before, vertically, 20 mm thick in two coats to brick base	19.1.3 The measurement is to the face of the wall. Rake out of joints of brickwork for key is deemed included.

				Basement 5
	19.51 0.10		Ditto level, width n.e. 150 mm, 20 mm thick in two coats to brick base	19.2.1
	4.74 4.74		Sand and cement (1:3) screed, beds and toppings >600 mm wide, level, 45 mm thick in one coat to asphalt base finished with a wood float	28.1.2
			&	
			Reinforced in situ concrete (20 N/mm² - 20 mm agg.) horizontal work, less than 300 mm thick in structures, bed	11.2.1 Lower bed
			x 0.15 = m³	
			Inner face 4000 Plus wall 2/490 980 4980 Less 2/102 ½ 205 4775 Less asphalt 2/2 20 40 4735	
4/	4.74		Formwork to edge of horizontal work, bed n.e. 500 m high, 150 mm high	
			&	
			Formwork to edge of horizontal work, suspended slab n.e. 500 mm high, 150 mm high	Suspended high-level slab

				Basement 6
	4.74 4.74 0.15		Reinforced in situ concrete ($20 N/mm^2$ - 20 mm agg.) horizontal work, less than 300 mm thick in structures, slab	11.2.1.2 Suspended slab
			Fabric reinf. 4735 Less cover 2/20 40 4695	
2/	4.70 4.70		Reinforcement fabric to BS 4483 type A142 weighing 2.22 kg/m^2 with 150 mm side and end laps	11.37.2
	4.00 4.00		Trowel surface of unset concrete to receive paving	11.8.1
			&	
			Formwork to soffit of slabs n.e. 300 mm thick, horizontal propping < 3.00 m high	11.15.1.1 Upper slab
			d.p.c. width Block 100 Cavity 75 Vertically 150 Brick 102 427	
			d.p.c. girth Face of brick 19 920 Wall 102 ½ Cav 75 Blk 100 Less 4/2/½/277 ½ 1 110 18 810	

				Basement 7
	18.81 0.43		d.p.c., exceeding 300 mm wide, horizontal pitch polymer cavity bedded in gauged mortar (1:1:6)	14.17.2.3
	27.40 0.10		Deduct Walls, half brick thick in commons as before	14.1.1.4
			&	
			Add Facing brickwork, walls, half brick thick, in stretcher bond, skins of hollow walls	14.1.1.1
	21.12 0.30 2.33		Deduct Disposal, excavated material off-site	5.9.2
			&	
			Add Filling to excavations exc. 500 mm thick with excavated material from site and compacting in 150 mm layers	5.11.2.2
	21.12 0.30 0.15		Filling to excavations, n.e. 500 mm thick with topsoil from spoil heap ave. 100 m distance, finished thickness 150 mm	5.11.1.1

Chapter 9
Walls

Brickwork thickness

As the thickness of the brickwork has to be given in the description, it is usually more convenient to give this in relation to the number of bricks. The generally accepted average size of a brick is 215 × 102.5 × 65 mm, which with a 10 mm mortar joint becomes 225 × 112.5 × 75 mm. This gives the following wall thicknesses:

- Half brick 102.5 mm
- One brick 215 mm
- One and a half brick 327.5 mm
- Two brick 440 mm.

Common and facing brickwork

When describing brickwork, it is necessary to distinguish between common brickwork and facework. Common brickwork is walling in ordinary or *stock* bricks without any special finish and usually hidden from view. If the brickwork is exposed to view and finished with a neat or fair face, then it is known as *facework*. Often, facework is built with a superior type of brick which gives a more pleasing appearance. Walls one brick thick and over may have facework on one side and common bricks on the other, this being a more economical way of constructing the wall if only one face shows. Facework is described as either one or both sides of the wall.

Willis's Elements of Quantity Surveying, Twelfth Edition. Sandra Lee, William Trench and Andrew Willis.

Published 2014 by John Wiley & Sons, Ltd.

Measurement of brickwork

Brickwork generally is measured as a superficial item, taking the mean girth of the wall (centre line) by the height. The heights of walls often vary within the same building, but it is usually possible to measure to some general level, such as the main eaves, and then to add for gables and so on, and deduct for lower eaves and the like. No deductions are made for voids within the area of brickwork not exceeding 0.50 m^2. In the case of internal walls and partitions, it is necessary to ascertain whether or not these go through floors and to measure the heights accordingly. If the thickness of a wall reduces, say, at floor level or at a parapet, then it should be remembered that the mean girth of the wall changes. All openings in walls are usually ignored when measuring at this stage, the deduction for these being made when the doors or windows are measured, as described in Chapter 13.

Sub-division

The measurement of walls can be conveniently divided into the following subdivisions:

- External walls
- Internal walls and partitions
- Projections of piers and chimney breasts
- Flues and chimney stacks.

NRM2 requires that brickwork has to be described in the following classifications:

- Walls
- Diaphragm walls
- Isolated piers and casings
- Attached projections
- Arches
- Bands
- Flues.

In each of these classifications, brickwork is described as vertical, battering, tapering (battered one side) or tapering (battered both sides). Isolated piers are defined as such when the length of the pier is less than four times its thickness, except where caused by openings.

Measurement of projections

Projections are defined as attached piers (if their length on plan is less than four times their thickness), plinths, oversailing courses and other similar items. Projections are measured linear, and the width and depth of the projection are given in the description. Horizontal, raking and vertical projections are kept separately. Example 4 at the end of this chapter gives sample measures for these.

Descriptions

The descriptions for brickwork, in addition to the matters discussed in this chapter, have to include the kind, quality and size of the bricks; the bond, composition and mix of mortar; and the type of pointing. These items, particularly if common to all the work, can be included in the bill as preambles or headings to reduce the length of descriptions. It should be remembered that, as mentioned here, mortar mixes may vary and, apart from substructure work, brickwork above eaves level in parapets and chimney stacks may well be specified to be in cement mortar.

Cutting, grooves and the like

All rough cutting on common brickwork and fair cutting on facework is deemed to be included. Rough and fair grooves, throats, mortises, chases, rebates, holes, stops and mitres are all also deemed to be included.

Returns and reveals

The labour to returns at reveals at the ends of walls is deemed included, and therefore nothing has to be measured for these.

Hollow walls

Each skin of a hollow wall is measured superficially and described as a wall, stating whether it has facework on one side. A superficial item is measured for forming the cavity, stating the width of the cavity.

The type and spacing of the wall ties have to be given in the description. If rigid sheet insulation is required in the cavity, this is also included in the description, stating the type, thickness and fixing method. Foam, fibre or bead cavity filling is measured separately as a superficial item, stating the width of the cavity, the type and quality of the material and the method of application. Filling cavities with concrete below ground level is measured as a cubic item, stating the thickness in the description as not exceeding 300 mm or as exceeding 300 mm. Closing cavities is measured linear, stating the width of the cavity, the method of closing and whether it is horizontal, raking or vertical.

Ornamental bands

These are items such as brick on edge or end bands, basket pattern bands, moulded or splayed plinth cappings, moulded string courses, moulded cornices and the like in facework. Horizontal, raking, vertical and curved bands are measured as separate linear items, stating the width and classified as follows:

- Flush
- Sunk (depth of set back stated)
- Projecting (depth of set forward stated).

If the bands are constructed entirely of stretchers or entirely of headers, this has to be stated, and if curved, the mean radius given. Ends and angles on the bands are deemed included.

Special purpose blocks or stones

The stones are numbered with a dimensioned description. The dimensions are taken over the stone plus one mortar bed and one mortar joint.

Reinforcement

Mesh reinforcement to brick joints is measured linear, stating the width and extent of laps. No allowance is made for laps in the measurement.

Measurement of arches

Arches are measured in meters along their centreline, stating the number of arches. The height of the face and width of the exposed soffit along with the shape of the arch are included in the description. Where arches are to be measured or the wall below deducted, the measurement is fairly straightforward for semicircular arches. In the case of segmental arches, the deduction above the springing line and the girth of the arch are not usually calculated precisely, as they can be estimated sufficiently accurately, the former from a triangle with compensating lines or by taking an average height, and the latter by stepping the girth round with dividers. In the case of expensive work, one should be as accurate as possible, the measurements preferably being worked out by calculation. A rough method of measuring the area of a segment is to take 11/16 times the area of the rectangle formed by the chord and the height of the segment. This is obviously not mathematically correct, and the margin of error will vary with the radius and length of chord, but when dealing with small areas this method will often be found sufficiently accurate. An alternative method is to take first the inscribed isosceles triangle based on the chord, leaving two much smaller segments, each of which may be scaled and set down as base × 2/3 height. The error is then very small. In dealing with large areas of expensive materials, a more accurate method would be necessary, using the formula:

$$\frac{H^3}{2C} + \left(\frac{2}{3}C \times H\right)$$

where C is the length of the chord and H the height.

Blockwork

The rules for measuring blockwork follow very closely those for measuring brickwork. Special blocks used, say, to achieve a designed bond at reveals, intersections and angles are measured as numbered items and described as extra over the work on which they occur.

Rendering

Rendering to external walls is usually taken with the walls and is measured the area in contact with the base as a superficial item. The description has to include the thickness, number of coats, mix and surface treatment, and the work has to be described as external. Rendering not exceeding 600 mm in width is measured as a linear item, and work to isolated columns is kept separately. Painting to the walls is measured as a superficial item and described as to general surfaces externally.

Stonework

The measurement of stone walling is considered to be beyond the scope of this book, but generally the rules follow those given for brickwork.

Internal partitions

These partitions are mentioned with this section so as to complete the measurement of the internal walls, although they are often measured with the internal finishes. When measuring timber stud partitions, all timbers including struts and noggins are measured linear and described as wall or partition members, stating their size in the description. All labours in the construction of the partition are deemed to be included. Door openings are usually measured net, as an adjustment later with the doors could be rather complicated. Plasterboard and other finishes are measured in square metres if over 600 mm wide, and lineally where under 600 mm.

Patent metal stud and other partitions are measured their mean length in square metres, stating their height in 1 m stages, the thickness and whether boarded one side or both sides. All sole and head plates, studs and the like are deemed to be included. Openings are numbered and measured as extra over the work in which they occur in 2.5 m² classifications. (See Fig. 16.)

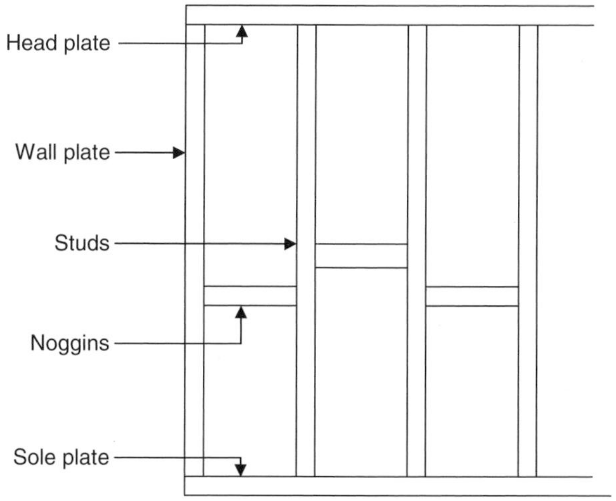

Fig. 16

Example 3

Walls and Partitions

(Above damp-proof course)

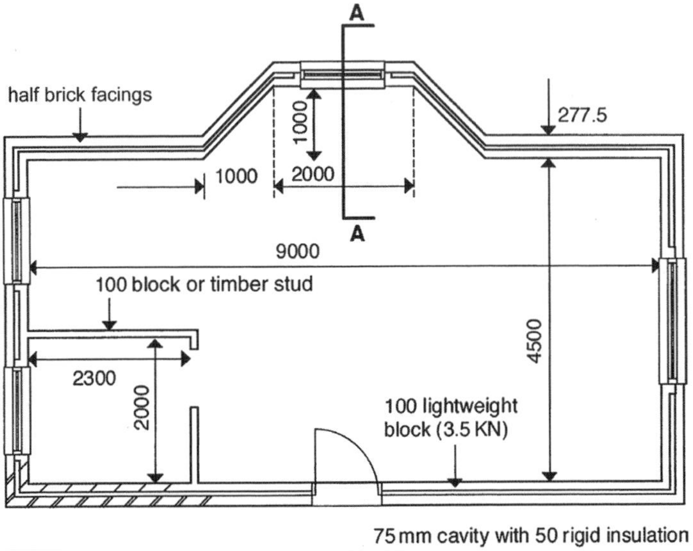

PLAN **Scale 1:100**

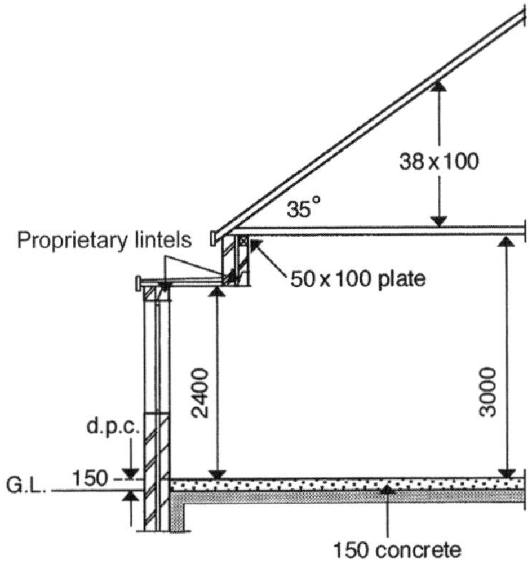

SECTION A-A **Scale 1:100**

Fig. 17

Example 3

Taking-off list	NRM2 reference
Cavity walls: block	14.1.2.1
facing brick	14.1.1.1
form cavity	14.14.1
Cavity insulation	14.15.1
Adjustment for bay: last four items	
Lintel	14.25.1
Gable ends	
Internal partitions	
Blocks	14.1.2.1
Stud partition	16.1.1.7

				External Walls 1
			<u>Calculation of mean girths excluding bay</u>	
			The approach taken here is to find the internal girth and to use the 4/2/the distance moved principle.	It is possible to adjust the internal girth to make allowance for the splay here; but in order to explain the calculation in detail, it has been taken separately on External Walls 3.
			9 000	
			4 500	
			2/13 000	
			Internal girth 27 000	
			$4/2/\frac{1}{2}/100$ 400	To centre of blocks
			Internal skin 27 400	
			27 000	
			$4/2/137\frac{1}{2}$ 1 100	To centre of cavity
			28 100	
			27 000	
			$4/2/226\frac{1}{2}$ 1 810	To centre of brick skin
			28 810	
			27 000	To centre of insulation
			4/2/125 1 000	
			28 000	
			<u>Height</u>	See Fig. 13 for an explanation of the distances moved in each calculation.
			Floor - ceiling 3 000	
			Less plate 50	
			2 950	

				External walls 2
	27.40 <u>2.95</u>		Dense concrete blocks, walls 100 mm thick in skins of hollow walls, cement mortar (1:4)	14.1.2.1
	28.81 <u>2.95</u>		Common brickwork, walls, half brick in stretcher bond, skins of hollow walls in cement mortar (1:4)	14.1.1.1
	28.10 <u>2.95</u>		Forming cavities in hollow wall, 75 mm wide, including stainless steel twisted wall ties, five per square metre	14.14.1
	28.00 <u>2.95</u>		50 mm thick cavity wall insulation, including polypropylene insulation retainers, five per square metre	14.15.1

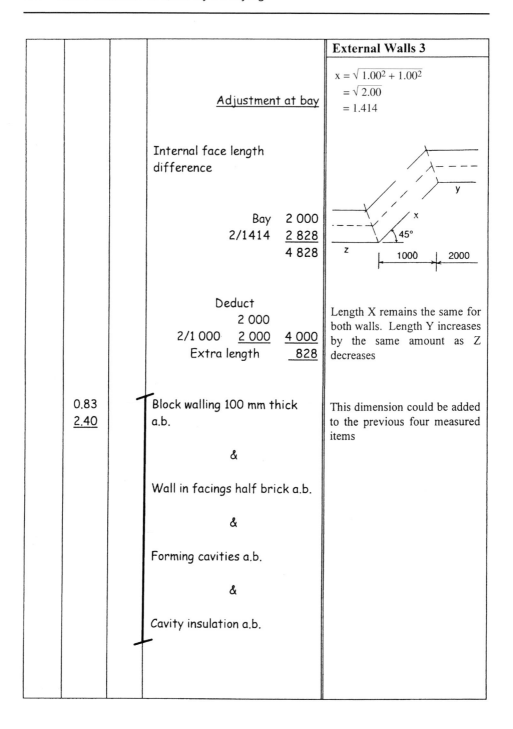

				External Walls 3
			Adjustment at bay	$x = \sqrt{1.00^2 + 1.00^2}$ $= \sqrt{2.00}$ $= 1.414$
			Internal face length difference Bay 2 000 2/1414 2 828 4 828	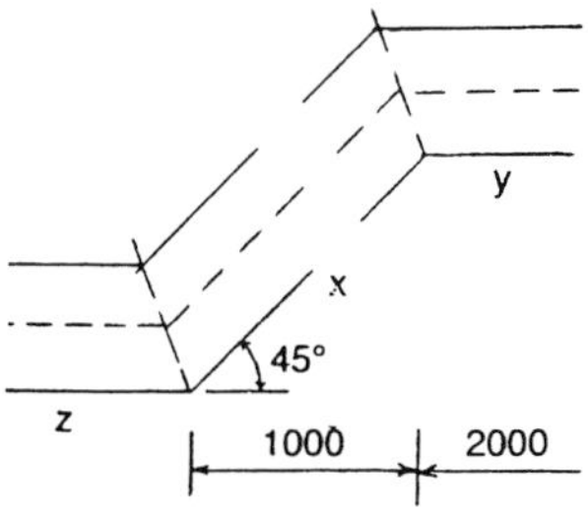
			Deduct 2 000 2/1 000 2 000 4 000 Extra length 828	Length X remains the same for both walls. Length Y increases by the same amount as Z decreases
	0.83 2.40		Block walling 100 mm thick a.b. & Wall in facings half brick a.b. & Forming cavities a.b. & Cavity insulation a.b.	This dimension could be added to the previous four measured items

				External Walls 4
	1		Proprietary lintel type x, 4.30 m long and building into brickwork	14.1.1 This is lintel across bay. Normally, lintels for windows would be taken with window measure.
			4 000 Bearing 2/150 300 4 300	
	4.30 0.23		Deduct Dense concrete blocks, walls 100 mm thick in skins of hollow walls, a.b.	14.1.2.1 Full courses only deducted
			ALTERNATIVE IF BUILDING GABLED ALONG TWO SHORT SIDES Blockwork 4 500 Plus 2/100 200 4 700 x $\frac{1}{2}$ = 2 350 2 350 x tan 35° Height of gable = 1 646	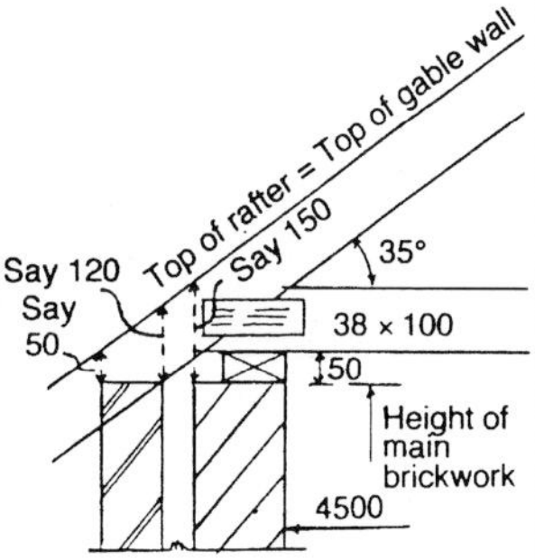 Extra height Plate 50 Rafter 100 150
2/	4.70 0.15		100 mm th block wall a.b.	14.1.2.1
2/$\frac{1}{2}$/	4.70 1.65			

				External Walls 5
			Brickwork	
			4 500	
			2/277½ 555	
			5 055	
			x ½ = 2 527½	
			2 527½ x tan 35°	
			Height = 1 770	
2/	5.06		Half brick wall in facings a.b.	
	0.05			
2/½/	5.06		Cavity	
	1.77			
			4 500	
			100	
			75	
			2/175 350	
			4 850	
			x ½ = 2 425	
			2 425 x tan 35°	
			Height = 1 698	
2/	4.85		Forming cavities in hollow walls a.b.	
	0.12			
2/½/	4.85			
	1.70			
2/	4.80		50 mm cavity insulation a.b.	
	0.14		4 500	
2/½/	4.80		100	
	1.68		50	
			2/150 300	
			4 800	
			x ½ = 2 400	
			2 400 x tan 35°	
			Height = 1 681	

Alternative stud partition example

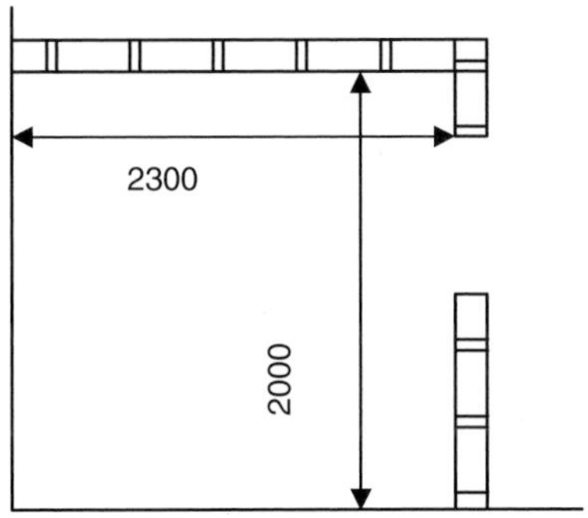

Fig. 18

A stud partition could be measured as an alternative to the block internal partition. It is assumed that 100 × 50 mm studs at 400 mm centres will be used. Head, cill and wall plates are also to be 100 × 50 mm sawn softwood. There would be wall plates fixed vertically at the junction with the inner block skin. Studs are then positioned at corners and openings to form the main structure. These might be a larger size than the basic studs or may even be double studs. Infill studs are then spaced equally between wall plates and corner studs or those at openings, at no more than 400 mm centres. Horizontal noggins are then required to brace the wall; in a 3 m high wall, it would be normal to have one line of horizontal members.

Adjustments for the door opening should be made after the members have been measured. To decide how many studs are required, the length of the wall is taken and divided by the spacings. The result must be rounded up to the nearest whole number and then, as this is the number of spacings, it is necessary to add one to allow for the end stud.

				Internal Walls 1
			<u>Lengths</u> 2 300 <u>2 000</u> 4 300 2/½/100 <u>100</u> <u>4 400</u>	
	4.40 <u>3.00</u>		Walls in lightweight concrete blocks (3.5 KN). vertical in gauged mortar (1:1:6)	
			ALTERNATIVE FOR STUD PARTITION 2 300 plus corner <u>100</u> 400) 2 400 = 6 + 1 = 7	The studs are 100 x 50 mm at 400 mm centres.
			400) <u>2 000</u> = 5 + 1 = 6	All adjustment will subsequently be made for the door opening.
			+ 1 Extra at corner total = 7 + 6 + 1 =14	
			<u>Height</u> 3 000 less 2/plates 2/50 <u>100</u> <u>2 900</u>	

				Internal Walls 2
2/	2.42		50 x 100 mm sawn treated softwood wall or partition members	16.1.1.7 Head and sole plates
2/	2.00			
14/	2.90		noggins	Studs
	3.70		2 400 2 000 4 400 less studs 14/50 700 3 700	
	0.90		Door adjustment Taken as size 900 x 2100 high	Door head
			height for deduction 2 100 plus head 50 2 150 noggins width 900 less vertical stud 50 850	
	2.15		Deduct last item	One vertical stud displaced by door and horizontal noggin
	0.85			

Example 4

Brick Projections

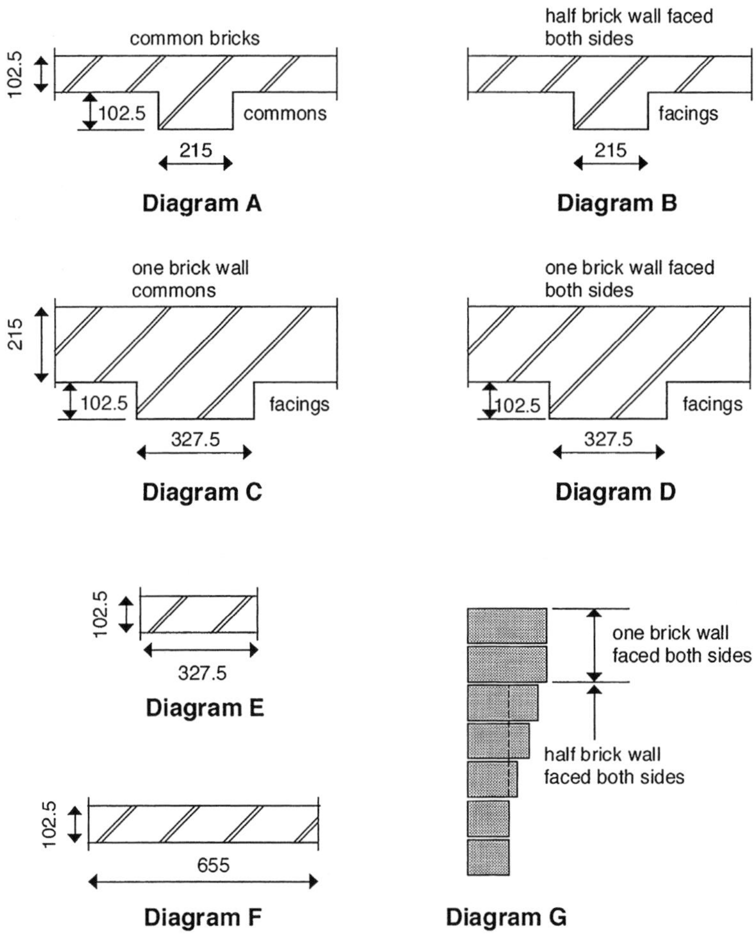

Plans: Diagrams A, B, C, D, E and F all 3000 high
Sections: Diagram G 5000 long, 3000 high

Scale 1:20

Fig. 19

Example 4

Taking-off list		**NRM2 reference**
Projections		
Diagram A		14.5.1
Diagram B	Adjust main wall	
	Projection	14.5.1
Diagram C	Adjust main wall	
	Projection	
Diagram D	Adjust main wall	
	Projection	14.5.1
Diagram E	Isolated pier	14.4.1
Diagram F	Wall	14.1.1
Diagram G	Adjust main wall	14.1.1
	Projection	14.5.3

				Brickwork 1
			Note: In the following examples, it has been assumed that the main wall has initially been measured ignoring the projection.	
			DIAGRAM A	
	3.00		Attached projections, in common bricks, one brick wide and half brick deep, vertical in gauged mortar (1:1:6)	14.5.1.1 Measured as an attached projection if the length on plan does not exceed four times the depth of projection.

				Brickwork 2
			DIAGRAM B	
	0.22 3.00		Deduct Walls half brick thick in facings, pointed both sides & Add Ditto but pointed one side	14.1.1 The original half brick wall in facings is substituted for the width of the pier by a wall pointed one side.
	3.00		Attached projection, in facing bricks, one brick wide and half brick deep, vertical in gauged mortar (1:1:6), including pointing with a weathered struck joint as the work proceeds.	14.5.1 Labours on returns ends and angles are deemed included.

				Brickwork 3
			DIAGRAM C	
	0.33 3.00		Deduct Walls one brick thick in commons, faced and pointed one side in gauged mortar (1:1:6)	The original wall faced one side is substituted with a one brick wall in commons (not face) for the width of the pier. 14.1.1
			&	
			Add Walls, on e brick thick in English bond in gauged mortar (1:1:6)	14.1.1
	3.00		Attached projection, facing bricks, in English bond, one and a half brick wide and half brick deep, vertical, in gauged mortar (1:1:6), including pointing with a weathered struck joint as the work proceeds.	14.5.1

				Brickwork 4
			DIAGRAM D	
			Deduct	
	0.33 3.00		Walls one brick thick in facings, pointed both sides a.b.	14.1.1 The original one brick wall in facings is substituted with one brick wall in common bricks, faced one side for the width of the pier.
			&	
			Add	
			Walls one brick thick in commons, English bond in gauged mortar (1:1:6), including pointing one side with a weathered struck joint as the work proceeds	14.1.1
	3.00		Attached projection, facing bricks, in English bond, one and a half brick wide and half brick deep, vertical, in gauged mortar (1:1:6), including pointing with a weathered struck joint as the work proceeds.	14.5.1 Full courses only deducted

				Brickwork 5
			DIAGRAM E	
	3.00		Isolated pier half brick thick vertical in common bricks, stretcher bond in gauged mortar (1:1:6)	14.4.1.1 Taken as isolated pier as length is less than four times the thickness of the wall.
			DIAGRAM F	
	0.66 3.00		Walls, half brick thick vertical in common bricks, stretcher bond in gauged mortar (1:1:6)	14.1.1 Length is more than four times the thickness; therefore, taken as a wall.

				Brickwork 6
			DIAGRAM G	
			2/75 150 3/75 225 375	Calculation of heights of brick courses
			Deduct	
	5.00 0.38		Walls half brick thick in facings, pointed both sides	The original wall has been measured for a height of 3 m and a length of 5 m, pointed both sides. This is now substituted with a wall pointed one side for the length of the oversailing courses.
			Add	
	5.00 0.23		Ditto but pointed one side	
	5.00 0.15		Ditto but one brick thick pointed both sides a.b.	
	5.00		Attached projection, facing bricks, average quarter brick thick and three courses high, horizontal, in gauged mortar (1:1:6), including pointing with a weathered struck joint as the work proceeds	14.5.3

Chapter 10
Floors

Timber sizes

Timber for use in construction is sawn into standard sectional dimensions, and the size thus created is known as the *nominal* or *basic* size. When the timber is processed, planed or wrot, it is reduced in size, usually by about 2 mm on each face, and the resultant size is known as the *finished* size. The original size of processed timber is sometimes described as *ex* or *out of*, thus 'ex 100 × 25 mm' would be '96 × 21 mm finished'.

Regularising is a machine process by which structural timber is sawn to a uniform size (e.g. joists sawn to an even depth). For regularising, 3 mm should be allowed off the size of timber of up to 150 mm, and 5 mm above this size.

NRM2 states that sizes of timber are deemed to be nominal unless described as finished. Confusion may be avoided if the sizes given in the bill are as they are shown on the drawings, provided that these are consistent. Structural timber when not exposed is usually left with a sawn finish, and the sizes given on the drawing are nominal. When architects are designing joinery, which may have to be to precise dimensions, they often prefer to give finished sizes. Care must be taken to ascertain which sizes are given on the drawing and to make it clear in the bill if finished sizes are being used.

The amount of timber removed in creating a wrot face on timber is known as the *planing margin*; this should be given as a preamble in the bill. Reference may be made to BS 4471, 'Dimensions for Softwood', which sets out in detail various planing margins according to the sizes and the end use of the timber.

Willis's Elements of Quantity Surveying, Twelfth Edition. Sandra Lee, William Trench and Andrew Willis.

Published 2014 by John Wiley & Sons, Ltd.

The specification for tongued and grooved floor boarding can cause confusion, as the finished width on face and the finished thickness may be quoted; for example, 90 × 21 finished boarding would be 100 × 25 nominal (sawn) size to allow for the tongue and planing margin.

Further consideration should be given to the sizes of timber readily available. For example, the width of a door lining to a 75 mm block partition with 13 mm plaster on both faces would require a finished lining width of 101 mm. The nearest nominal size of timber available is 115 mm, or 125 mm for some types of timber.

Timber exceeding 5.1 m in length is usually more expensive; in the examples, an allowance in the length of timbers has been made for jointing where appropriate. NRM2 requires structural timbers over 6 m long in one continuous length to be measured as such (16.1.1.1.9).

Subdivision

The measurement of floors is conveniently divided into two main subdivisions of finishes and construction. These subdivisions may be taken storey by storey, or the finishes may be taken first for the whole building, followed by the construction. Whether or not the finishes are taken before the construction is really a matter of personal choice, although measurement, and therefore knowledge, of the finishes may assist in deciding the form of construction to be used. For example, a timber upper floor may have a general finish of boarding, but a small tiled area may require a different form of construction.

Frequently, floor finishes are measured with the internal finishes section, because measurements used for ceilings may apply to floors. This is, however, a matter for agreement between the surveyors measuring the two sections, although it is customary to measure the finish to timber floors with the construction.

Timber floor construction

The measurement of timber floors may be divided as follows:

- Plates and/or beams
- Joists
- Hangers
- Strutting
- Ties to walls
- Insulation.

Timber-suspended ground floors are not usually cost effective but may be used to avoid excessive fill below solid floors or on sloping ground. The under-floor space should be ventilated; air bricks and sleeper walls built honeycombed with a damp-proof course if required, and they are measured with either floors or substructure but billed with the latter.

Timber plates and bearers are measured linear, and the length measured should allow for any laps or joints required. Timber joists are measured in the same way but are described as floor joists.

To ascertain the number of joists in a room, take the length of the room at right angles to the span of the joists and subtract, say, a 25 mm clearance plus half the thickness of the joist at each end. This will give the distance from the centre lines of the first and last joists, which is divided by the spacing of the joists. The result, which should be rounded to the next whole number, gives the number of spaces between the joists. To this number, one must be added to give the number of joists rather than the spaces. Two points should be borne in mind before making the calculation. Firstly, the room size used for the calculation should be that of the room below the floor as the joists will be supported by the walls of that lower room. Secondly, if any of the intermediate joists are to be in a fixed position, such as a trimmer around an opening, two separate dimensions should be taken for the division on either side of the fixed joists, thus avoiding any increase in the maximum spacing. Additional joists may be required to support upper floor partitions, and care must be taken to include for these. Fig. 20 shows how to calculate the number of joists required for a timber ground floor.

The trimming of timber joists for staircases, ducts, hearths, access panels and so on is usually taken with the floor construction. Joists used in trimming should be increased in thickness by 25 mm, and an addition made to their length for jointing unless metal hangers and framing brackets are used. Displaced joists and floor coverings have to be deducted.

Actions to prevent joists twisting, herringbone or solid strutting, which may not be shown on the drawings, should be taken. This is measured linear over the joists, and there should be one row, say, every 2.4 m.

Building regulations may require ties to be provided where timber joists run parallel to an envelope wall; these are probably best taken with the floor construction.

Staircases

Timber staircases are enumerated and may be either described fully with dimensions or accompanied by a component detail. Items such as linings, nosings, cover moulds, trims, soffit linings, spandrel panels, ironmongery,

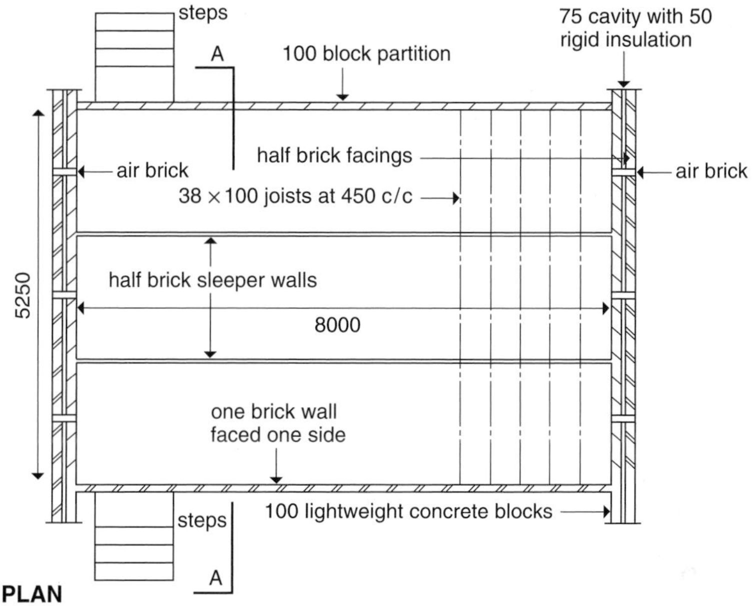

PLAN

			Joists 8 000 Clearance 25 $\frac{1}{2}$ Joist 19 - 2/44 88 ÷ 450)7 912 18 + 1 Number of joists = 19	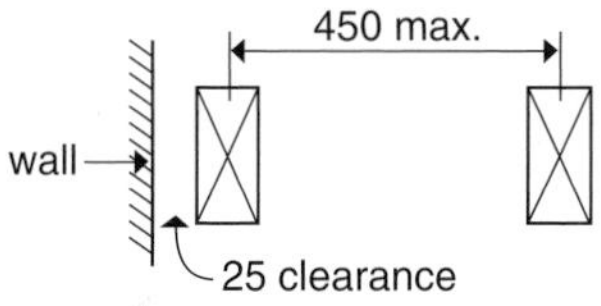 The nearest whole number is taken and one added to the number of spaces.

Fig. 20

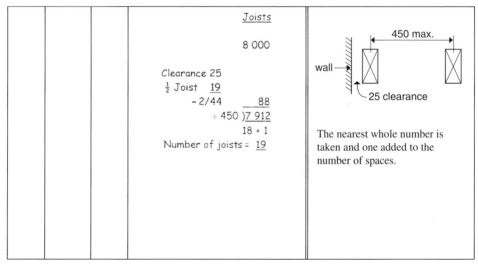

finishes, fixings, wedges and the like, where supplied with or with part of the component, are deemed to be included. They would, of course, have to be included in the description or shown on the component detail. If these items are not part of the composite item, then they are measured separately according to the appropriate rules. For example, cover fillets and nosings are measured as linear items with ends being deemed included. The creation of the stairwell is usually measured with the floors section, but it may be necessary to adjust the internal finishes at this stage. Decoration on the staircase itself, if not carried out at the factory, has to be measured. Handrails and balustrades that are isolated and do not form part of a staircase unit are measured as linear items, and their size is stated. Ramps, wreaths, bends and ornamental ends on handrails are deemed included.

Concrete floors

The measurement of concrete upper floors is described in Chapter 14 as part of reinforced concrete structures.

Precast concrete beam and pot floors

A common ground floor construction used in UK domestic properties comprises a series of precast concrete ribs (with an inverted T shape) supporting concrete blocks. This normally rests on block walls. The floor is measured as a composite item, including the ribs and beams as described in Fig. 21 and the sample form in Example 5.

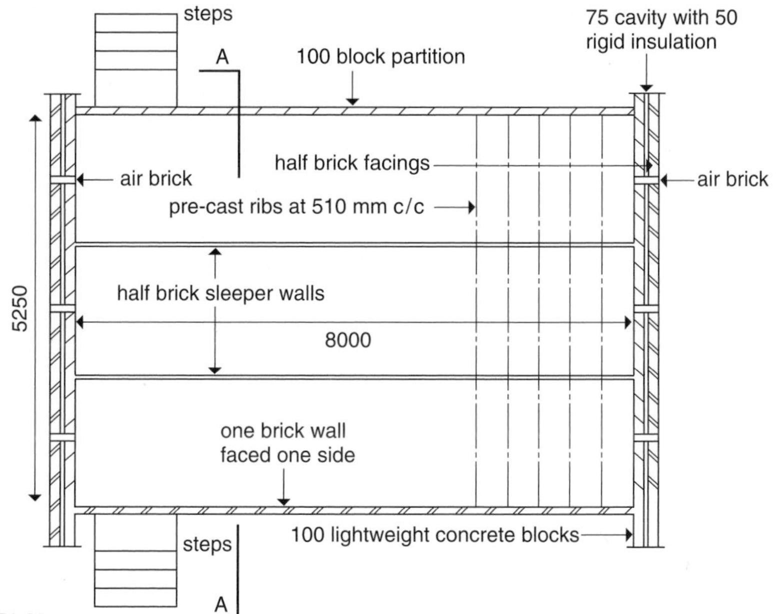

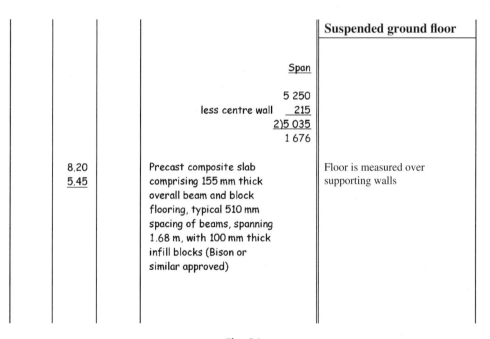

				Suspended ground floor
			<u>Span</u> 5 250 less centre wall <u>215</u> <u>2)5 035</u> 1 676	
	8.20 <u>5.45</u>		Precast composite slab comprising 155 mm thick overall beam and block flooring, typical 510 mm spacing of beams, spanning 1.68 m, with 100 mm thick infill blocks (Bison or similar approved)	Floor is measured over supporting walls

Fig. 21

Example 5

Timber Upper Floor

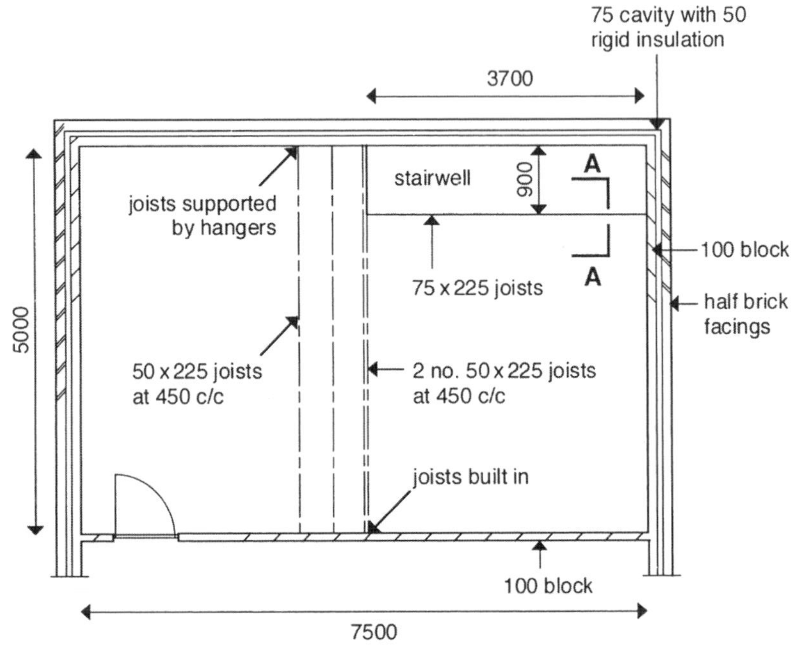

PLAN **Scale 1:100**

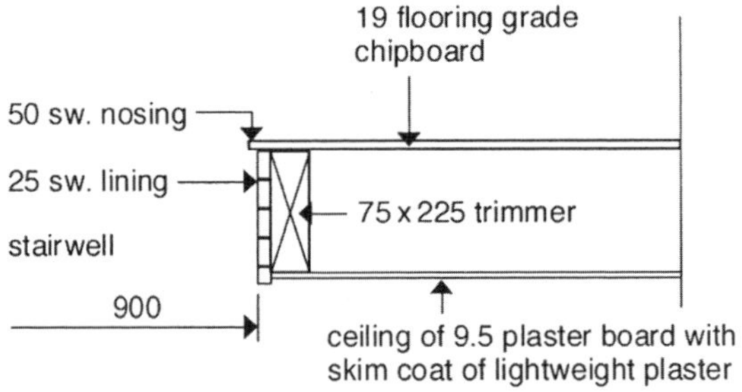

SECTION A–A **Scale 1:20**

Fig. 22

Example 5

Taking-off list	NRM2 reference
Floor joists	16.1.1.4
Trimmer joists	
Adjust at stairwell	
Joist hangers	16.6.1.7
Herringbone strutting	16.1.1.8
Steel straps	16.6.1.4
Floor boarding	16.4.2.1
Nosings	16.4.1.3
Apron lining	22.3.1
Decoration to lining	29.1.1.1

				Timber Upper Floors 1
			3 700 Less clearance 25 Joist ½/50 = 50 450)3 650 = 9 7 500 - 3 700 3 800 As above 50 2/50 100 - 150 450)3 650 = 9 Length 5 000 + 100 5 100	Where joists are in a fixed position, the number of joists should be calculated from that position. In the example, the stair trimming joist is used as the base.
2/9/ 2/	5.10 5.10		Floor members 50 x 225 mm treated sawn softwood (double trimming joist)	16.1.1.4
			900 75 975	
9/	0.98		Deduct Last item	Adjustment of joists in stairwell
	3.70		Floor members, 75 x 225 mm a.b.	Stair trimmer

				Timber Upper Floors 2
9/ 2/	1 1		3 mm welded mild steel galvanised joist hangers for 50 x 225 mm members to blockwork	16.6.1.7
2/	1		Ditto for 75 x 225 mm member	For trimmer joist
9/	1		Ditto for 50 x 225 mm, but fixed to timber	For ends of trimmed joist
	7.50		Herringbone strutting 38 x 19 mm, treated sawn softwood to 225 mm deep joists	16.1.1.8 Taken to give side support to joists
2/2/	1		Galvanised mild steel joist restraining strap 30 x 5 x 1000 mm girth, once bent and 3 x countersunk drilled.	16.6.1.4 These should be taken every 2 m where joists are parallel to the external wall. Fixing deemed included.

				Timber Upper Floors 3
	7.50 5.00		19 mm flooring grade chipboard exceeding 300 mm wide 3 700 900 0 10 3 710 910	16.4.2.1
	3.71 0.91		Deduct Last item	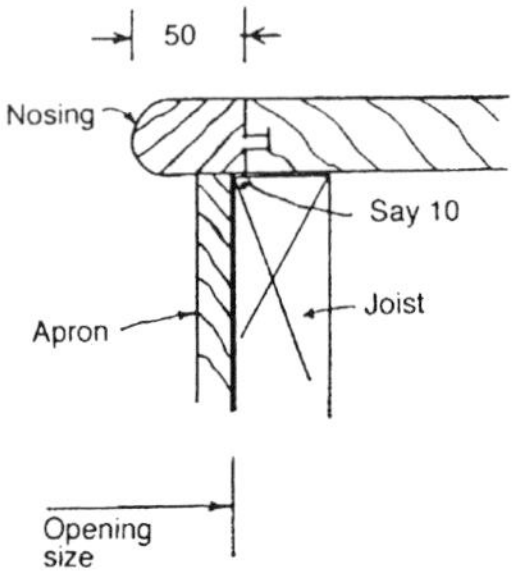
	0.91 3.71		Nosing 50 x 25 mm wrot softwood, twice rounded, tongued to edge of chipboard including groove.	16.4.1.3
	3.70 0.90		Apron lining 25 x 250 mm wrot softwood 225 Ceiling 9 ½ Clearance 10 244 ½	22.3.1
			&	
	3.00		Knot, prime and stop and two undercoats and one gloss coat to general surfaces n.e. 300 mm girth in staircase areas.	29.1.1.1

Chapter 11
Roofs

Subdivision

The measurement of roofs is subdivided conveniently into two main sections on coverings and construction. As with floors, if the coverings are measured first, it should be easier to understand the construction. However, it is common practice to measure the construction first, starting with the wall plate and following the order in which the roof would be built. When a building has different kinds of roof (e.g. tiled roofs and felt flat roofs), each should be dealt with separately. If there are several roofs of similar type, it may be more convenient to group these together. The rainwater system, unless designed with the plumbing installation, is measured with the roofs section, and may be taken either at the end of each type of roof or at the end of the entire roof measurement.

If no roof plan is provided, this may be superimposed on the plan of the upper floor. Special care must be taken to ensure that all hips and valleys are shown, one or the other being necessary at each change of direction in a pitched roof. In the case of a flat roof, it is necessary to show the direction of fall, gutters and outlets.

Pitched roof construction

The notes on timber sizes and the measurement of timber floor construction in Chapter 10 should be understood before reading this section. With traditional timber roof construction, a logical sequence similar to the following should be adopted:

Willis's Elements of Quantity Surveying, Twelfth Edition. Sandra Lee, William Trench and Andrew Willis.

Published 2014 by John Wiley & Sons, Ltd.

	NRM2 reference
Plates	16.1.1.3
Rafters	16.1.1.1
Ridge	16.1.1.1
Hips and valleys	16.1.1.1
Ceiling joists and collars	16.1.1.1
Purlins, ceiling beams, hangers and struts	16.1.1.2
Sprockets	16.1.1.1
Adjustments for openings, dormer construction and so on (as above)	
Ties to walls	16.6.1.1
Walking boards	16.4.1.1
Insulation	31.3.1.3
Ventilation	16.4.1.1

The number of rafters in a gabled roof should be calculated by taking the total length of the roof between the centre lines of the end rafters and dividing this length by the rafter spacing. The result should be rounded to the next whole number, and one added to convert the number of spaces to the number of rafters. The number of rafters will have to be multiplied by two for the two slopes. If an intermediate rafter such as a trimmer is in a fixed position, then the calculation should be made on either side of the fixed rafter. If the end of the roof is hipped, then a similar calculation can be made taking the dimension from the foot of the jack rafters at the hipped end. One extra rafter should be taken at each hipped end opposite the ridge as shown in Fig. 23. It will be seen that the jack rafters at the hipped end are equal to the lengths of the common rafters (shown dotted) that would have been there if the roof were gabled. The number of ceiling joists may be calculated in a similar way, but the length to be divided is taken between the inside face of the external walls less the clearance as shown for timber floor construction.

All structural timber in pitched roofs is measured linear and described as in pitched roofs, stating the size in the description. Plates and bearers are kept separately and described as plates. Metal connectors, straps, hangers, shoes, nail plates, bolts and metal bracing are enumerated.

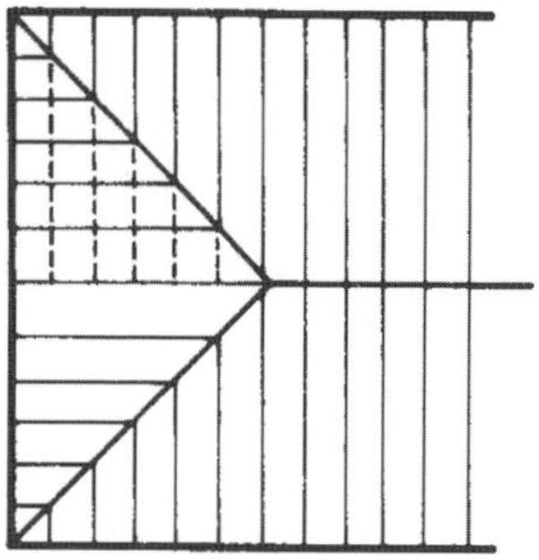

Fig. 23

Roof slopes

The roof slope of a pitched roof is the hypotenuse of a triangle, the base of which is half the roof span and the height that of the roof. In the triangle ABC in Fig. 24, AB could be found by using Pythagoras if the lengths AC and CB were known. It is more common for the span of a roof and the pitch to be given on the detail drawings.

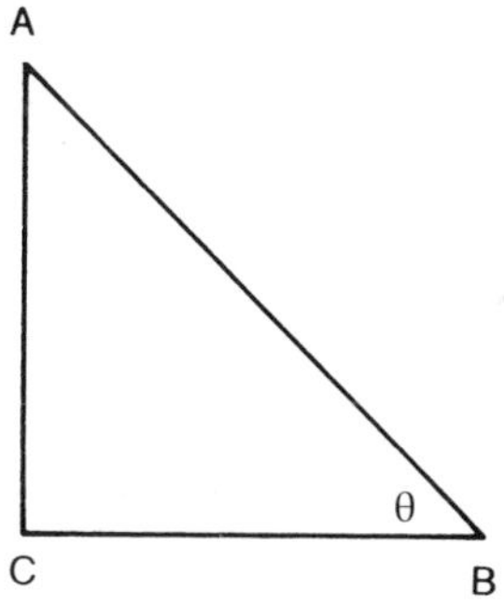

Fig. 24

Here we would use the following formula:

$$\text{cosine}\theta = \frac{BC}{AB}$$

therefore:

$$AB = \frac{BC}{\text{cosine}\theta}$$

Consider the example in Fig. 25. Half the roof span must of course be calculated to the extreme projection of the eaves or the edge of the tiling

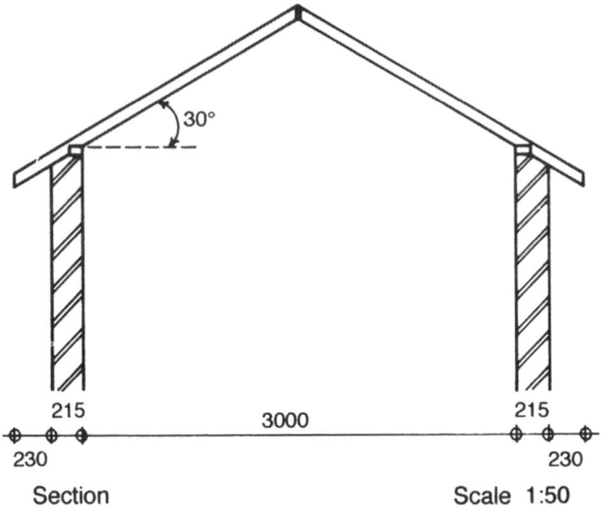

Fig. 25

if coverings are being measured. The span inside the walls being 3000 mm, half the span for this purpose will be 1500 + 215 + 230 = 1945 mm. The angle of pitch being 30°, the length of the slope will be:

$$
\begin{aligned}
& 1945 \div \text{cosine } 30^\circ \\
&= 1945 \div 0.866 \\
&= 2246 \text{ mm}
\end{aligned}
$$

and this can be checked by scaling.

It should be noted that this calculation gives the length along the top of the rafter to the centre line of the ridge. This may have to be adjusted for the length of the covering, which may project into a gutter.

Hips and valleys

The length of a hip or valley in a pitched roof must be calculated from a triangle, there usually being no true section through it from which it can be scaled. When the pitch of the hipped end is the same as that of the main roof, then the length of the hip may be found from half the span and the length of the roof slope using Pythagoras.

For example in Fig. 26:

$$\text{BE and BD} = \frac{1}{2}\text{span}$$

$$\text{AD and AE} = \text{length of slope}$$

$$\therefore \text{AB} = \sqrt{\text{AD}^2 + \text{BD}^2}$$

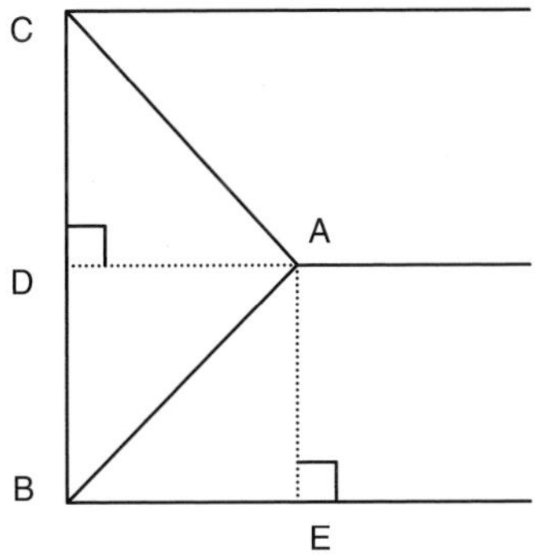

Fig. 26

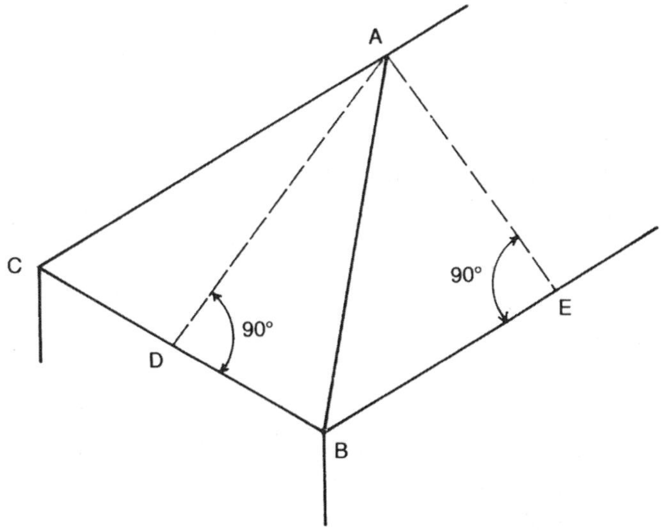

Fig. 27

Broken-up roofs

It should be noted that when a roof is broken up by hips and valleys, so long as the pitch is constant the area will always be the overall length multiplied by twice the slope. The area of tiling on a roof hipped at both ends is therefore measured in the same way as if it were gabled, the only difference being that if it were gabled the dimension of the length would probably be smaller, the projection of verges being less than that of eaves.

For example, in Fig. 28:

Area of triangle ABD = area of triangle ABF

$$BE = BD = AF = \frac{1}{2}\text{ span}$$

AB is common

$A\hat{D}B = 90°$ and $A\hat{F}B = 90°$

With two sides and one angle equal, the triangles ABF and ABD are equal in area.

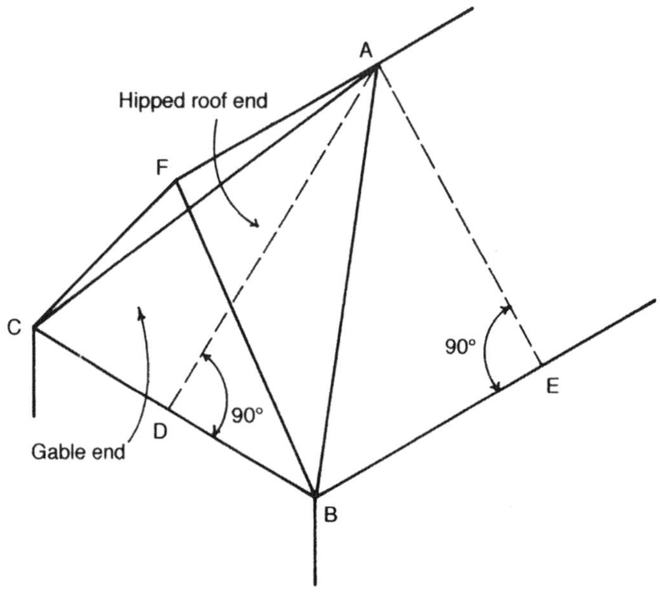

Fig. 28

Just as the length of slope is the length on plan divided by the cosine of the angle of pitch, so the area of a roof of constant pitch is the area on plan divided by the cosine of the angle of pitch; this formula can be a useful check on the measurements when worked out.

Trussed rafters

Trussed rafters are enumerated and described, and although there is no specific requirement in NRM2 for a dimensioned diagram to be given, this is often the clearest way to describe a truss. Bracing and binders to the construction are taken as linear items. When calculating the number of trussed rafters, consideration should be given to the position of water storage tanks and the like, as their support may require additional trusses, platforms and bracing. This will probably involve discussion with the surveyor measuring the plumbing installation; agreement should be reached on where additional items will be measured and by whom.

Tile or slate roof coverings

The first measurement in this type of roof should be the area of coverings; the description will include battens and underlay. If the area of one slope is entered in the dimensions, care must be taken to multiply this by two to allow for both slopes. Following the coverings, one should proceed around the edges of the roof, measuring abutments, eaves, verges, ridges, hips and valleys. The adjustments for chimneys and dormers and any appropriate metal gutters, soakers and flashings could conclude the coverings measurement or be taken as a separate section at the end after the construction.

Unless otherwise specified, customary girths for leadwork to pitched roofs are as follows:

• Cover flashings	150 mm
• Stepped flashings (over soakers)	200 mm
• Ditto (without soakers)	300–350 mm
• Aprons	300 mm
• Soakers 200 mm wide	length = gauge + lap + 25 mm (number = length of slope ÷ gauge).

Eaves and verge finish

Verge, eaves, fascia and barge boards are measured linear, giving their size in the description. Painting the boards should be taken at the same time as a superficial item unless the member is isolated and less than 300 mm wide, when it is measured linear.

Rainwater installation

Rainwater gutters are measured linear over all fittings, these being enumerated and described as extra over the gutter. Rainwater pipes are measured in a similar manner and, in addition, stating whether they are fixed in ducts, chases, screeds or concrete. The description of gutters and rainwater pipes should include their size, method of jointing, type of fixing and whether to special backgrounds. Rainwater heads are enumerated and gratings may be included in the description.

A careful check should be made to ensure that all roof slopes are drained, that there are sufficient rainwater pipes and that the drainage plan shows provision for each pipe. If the sizes of gutters and pipes are not shown then these have to be calculated, but it should be remembered that a variety of sizes, although theoretically correct, does not always lead to an economic solution.

Flat roofs

The same main subdivision of coverings and construction can be made as before. The examples show the measurement of asphalt flat roofs. The measurement of metal covered roofs should not prove too difficult provided that their construction is fully understood.

Metal roofs are measured their net visible area with extra over linear items measured for drips, welts, rolls, seams, laps and upstands. No deductions are made for voids not exceeding 1 m^2 within the area of the roof. When the opening has not been deducted, no further work around the opening has to be measured except in the case of holes. The pitch of the roof has to be stated in the description.

Flashings are measured linear, stating the girth; dressing into grooves, ends, angles and intersections is deemed included. Gutters are measured in a similar manner and labours are deemed to be included. Outlets are enumerated, stating their size, and again labours are deemed included.

Asphalt roofing is measured the area in contact with the base and again no deductions are made for voids within the area which do not exceed 1 m^2. The pitch has to be stated in the description. The width of work has to be classified as not exceeding 500 mm or as exceeding 500 mm. Skirtings, fascias, aprons, channel and gutter linings, and coverings to kerbs are measured linear the exact girth being given. All labours to these items are deemed included.

Felt roof coverings are measured superficial, stating the pitch. Skirtings, gutters, coverings to kerbs and so on are measured linear.

Timber flat roof construction is measured in a similar manner to that of floors. The fall in the roof may be created by the use of firrings, which are measured linear, stating their thickness and mean depth. Gutter boards and sides are measured superficial, classifying the width as exceeding or not exceeding 600 mm. Timber drips and rolls, if required for joints in metal coverings, are measured linear, stating their size. Fascias, soffits and rainwater goods are taken in a similar manner to those for pitched roofs.

Concrete roofs are measured in a similar manner to floors as described in Chapter 14. The fall in the roof is usually created in the screed; however, if the slab is sloping, then this is classified as either exceeding or not exceeding 15°. If the slab exceeds 15°, formwork has to be measured to the upper surface. Concrete upstands are measured as cubic items; formwork to the sides is measured in square metres. Trowelling the surface of the concrete is measured as a superficial item.

Example 6

Traditional Pitched Roof

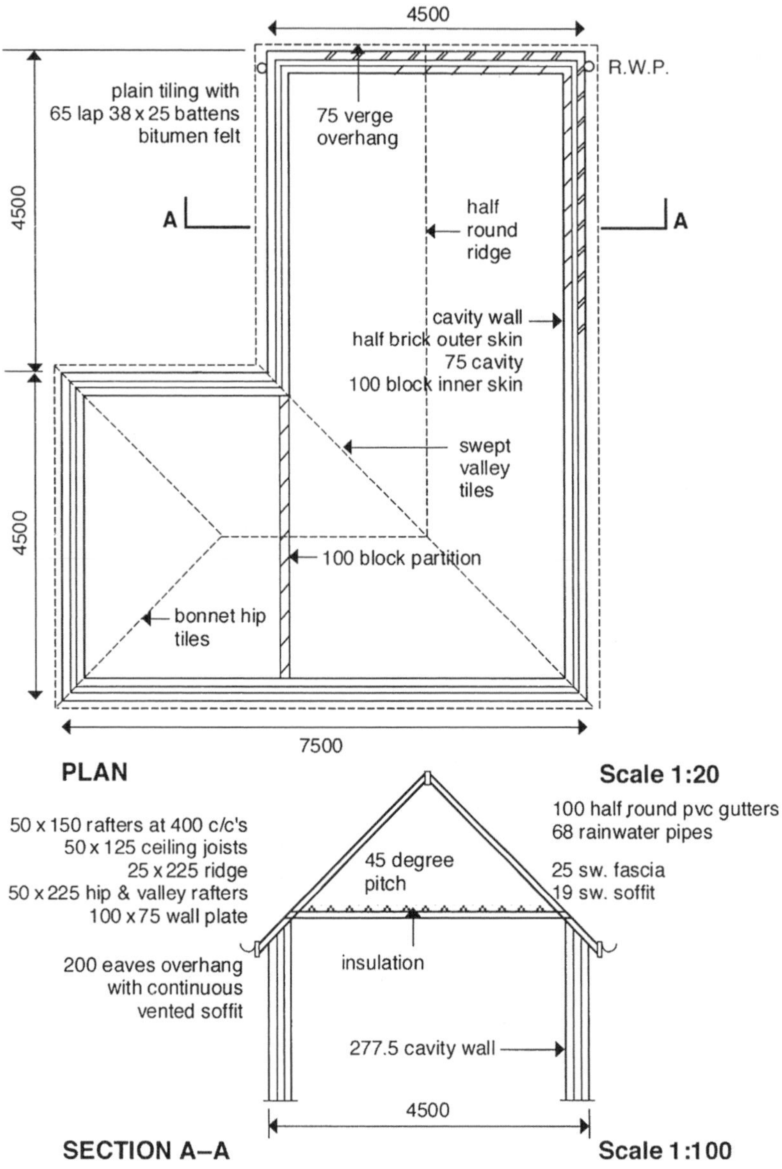

Fig. 29

Example 6

Taking-off list	NRM2 reference
Structure	
Wall plates	16.1.1.3
Roof members: rafters	16.1.1.1
ridge, hip, valley	
ceiling joists	
Coverings	
Tiling, battens and underlay	18.1.1.1
Ridge	18.3.1.3
Saddle	18.4.1.6
Hip tiles	18.3.1.6
Valley tiles	18.3.1.5
Verge tiles	18.3.1.4
Eaves course	18.3.1.2
Tilting fillet	16.3.1.1
Fascia	16.4.1.3
Soffit board	16.4.1.5
Battens	16.3.1.1
Spandril ends	16.3.1.6
Decoration to eaves	29.1.2.2
Eaves ventilator	16.4.1.1
Insulation	31.3.1.3
Rainwater goods	
Gutters	33.5.1.1
Fittings	33.6.1.1
Down pipes	33.1.1.1
Fittings	33.3.1.4

Traditional Roof 1

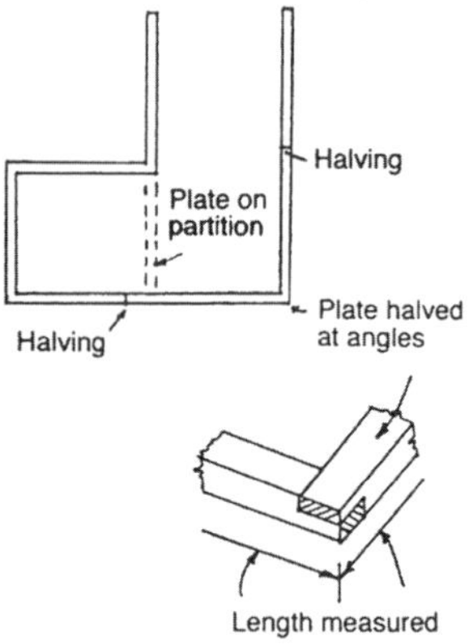

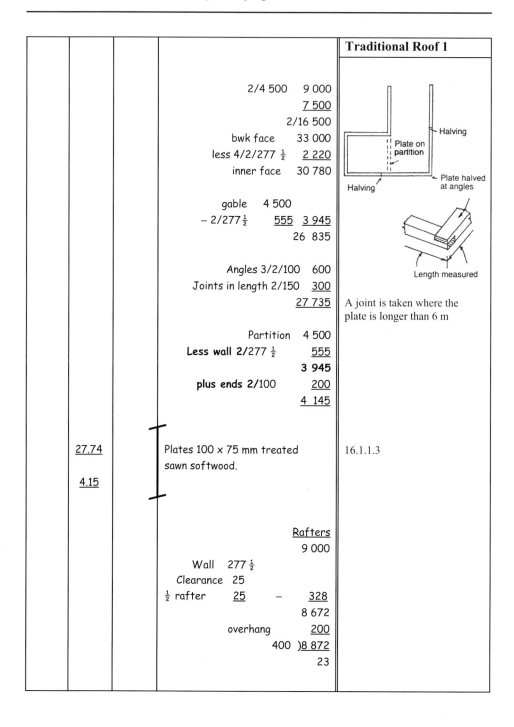

			2/4 500 9 000	
			7 500	
			2/16 500	
			bwk face 33 000	
			less 4/2/277 ½ 2 220	
			inner face 30 780	
			gable 4 500	
			– 2/277½ 555 3 945	
			26 835	
			Angles 3/2/100 600	
			Joints in length 2/150 300	
			27 735	A joint is taken where the plate is longer than 6 m
			Partition 4 500	
			Less wall 2/277 ½ 555	
			3 945	
			plus ends 2/100 200	
			4 145	
	27.74		Plates 100 x 75 mm treated sawn softwood.	16.1.1.3
	4.15			
			Rafters	
			9 000	
			Wall 277 ½	
			Clearance 25	
			½ rafter 25 – 328	
			8 672	
			overhang 200	
			400)8 872	
			23	

				Traditional Roof 2
			7 500 – 4 500 400)3 000 = 8 Rafter length 2 250 200 2 450 Fascia 25 $\frac{1}{2}$ Ridge 12 $\frac{1}{2}$ – 37$\frac{1}{2}$ 2 412 $\frac{1}{2}$ Cos 45^{C} = 3 411 $\frac{1}{2}$/150 75 Rafter length 3 486	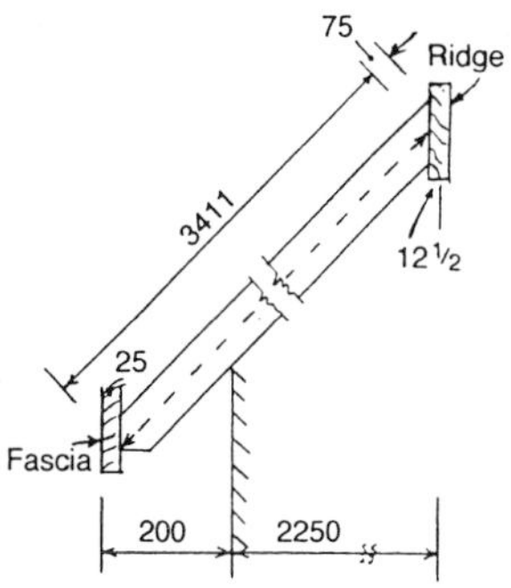
2/23/	3.49		Structural timbers, rafters and associated roof members 50 x 125 mm treated sawn softwood Ridge 7 500 – $\frac{1}{2}$/4 500 2 250 5 250 4 500 9 750 Gable brick 102 $\frac{1}{2}$ cavity 75 – 177 $\frac{1}{2}$ 9 572$\frac{1}{2}$ Joint 2/225 450 10 022$\frac{1}{2}$	G20.9.2.1 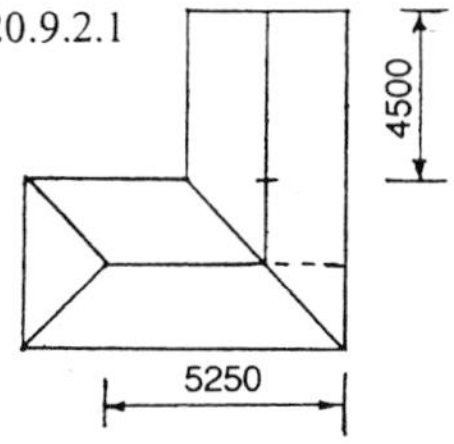 16.1.1.1 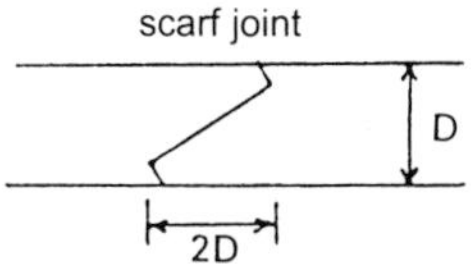 Length over 6 m
	10.02		Structural timbers, rafters and associated roof members 25 x 225 mm treated sawn softwood	16.1.1.1

				Traditional Roof 3
			$3\,486^2 + 2\,4112^2 = Hip^2$ Hip = 4 239	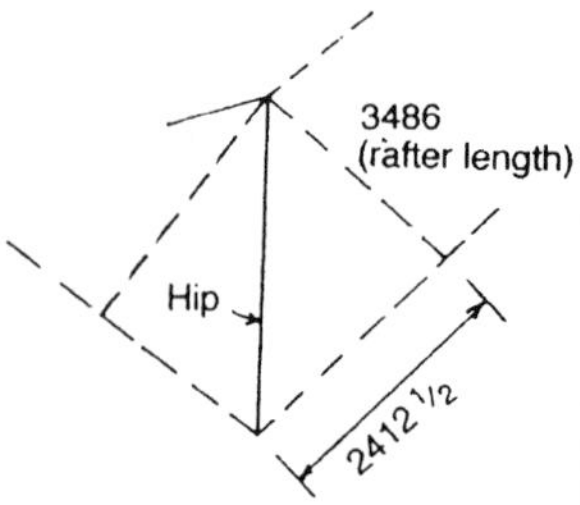
4/	4.24		Structural timbers, rafters and associated roof members 50 x 225 mm treated sawn softwood	16.1.1.1
			Ceiling joists	
			9 000 Wall 277 ½ Clearance 25 ½ Joist = ½ 50 25 – 2/377 ½ – 655 400)8 345 = 21 + 1 = 22	
			7 500 – 4 500 3 000 – 327½ 2 672½	
			107½ 75 177½ 2 850	
			25 ½/50 25 – 50 400)2 800 = 7 + 1 = 8	

				Traditional Roof 4
30/	4.30		Structural timbers, rafters and associated roof members, 50 x 125 mm treated sown softwood	16.1.1.1 Ceiling joists: 22 + 8 = 30.
			4 500 2/327 ½ 655 400) 3 845 = 10 + 1 = 11	Where joists rest on partition at junction of L shape.
11/	0.15		Deduct Last item	
			Coverings 7 500 Overhang 200 Into gutter 50 2/250 500 8 000 plus return 4 500 length = 12 500 slope length 4 500 eaves a.b. 500 2)5 000 ½ span = 2 500 Slope = 2 500 / Cos θ = 3.535	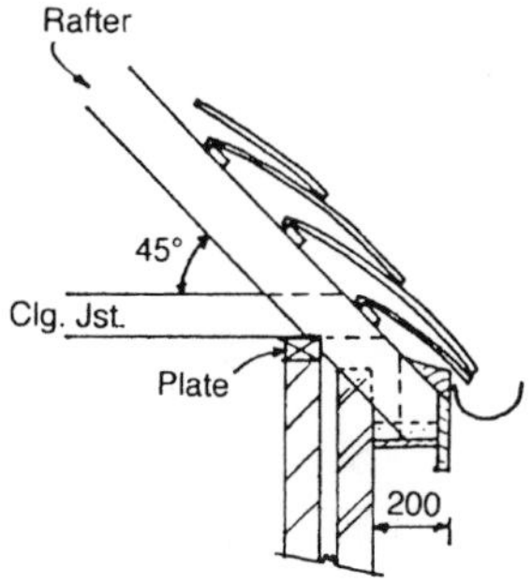

				Traditional Roof 5
2/	12.50 3.54		Roof coverings, 45° pitch, handmade sand-faced plain clay, tiles size 265 x 165 mm, 65 mm end lap, nailed every fourth course with two aluminium nails, 38 x 25 mm treated softwood battens with galvanised nails and reinforced bitumen felt to BS747 type IF weighing 15 kg/m^2 with 75 mm horizontal and 150 mm end laps	18.1.1.1 The area of the roof coverings does not need to be adjusted for the hipped end, as long as the pitch on the hipped end is the same as the main slope.
			Ridge 2/4 500 9 000 Eaves 250 Verge 75 325 9 325 + return 8 000 17 325 – 3/½/5 000 7 500 9 825	
	9.83		Ridge 250 mm diameter **half** round, red clayware	18.3.1.3
2/	1		Saddle 450 x 450 mm Code 4 lead, supply only & Fix only ditto	18.4.1.6 At angle of ridge and junction with hips

				Traditional Roof 6
			Hips Using Pythogoras, the hip length is √(3.535 ^{2} + 2 500^2) = 4 330	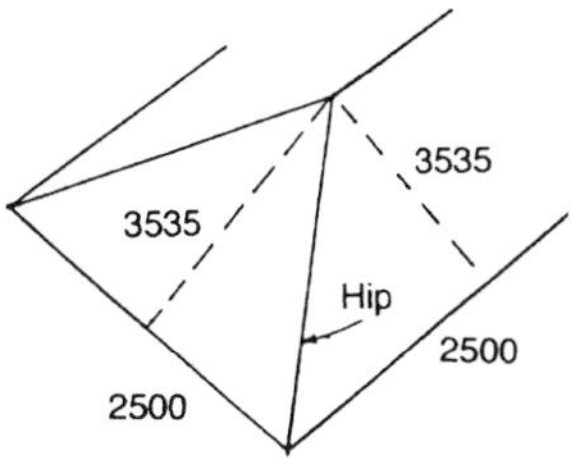
3/	4.33		Hip, matching bonnet hip tiles	18.3.1.6
	4.33		Purpose-made swept valley tiles	18.3.1.5
	3.54		Verge with plain tile undercloak	18.3.1.4 Undercloaks deemed included but method of forming to be stated.
			Eaves 8 000 9 325 2/17 325 34 650 Less gable 4 500 Eaves 2/250 500 5 000 29 650	
	29.65		Eaves double tile course	18.3.1.2
			Tilting fillet 29 650 Less overhang 2/2/50 200 29 450	
	29.45		Battens 75 x 50 mm extreme treated softwood triangular	16.3.1.1

				Traditional Roof 7
	29.45		Fascia board 25 x 175 mm wrot softwood, chamfered and grooved	16.4.1.3
			Soffit Fascia length 29 450 + for int. corner 2/200 400 29 850	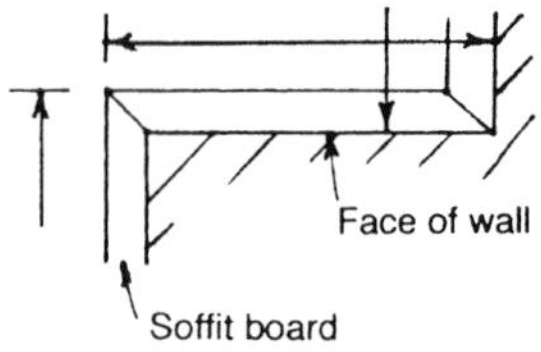
	29.85		Eaves soffit board, 19 x 200 mm wrot softwood, rebated	16.4.1.5
			Bwk face 33 000 Less gable 4 500 400)28 500 = 72 +1 = 73	Brackets at rafter position for soffit fixing
73/	0.20		Battens 38 x 50 mm treated sawn softwood	Horizontal
73/	0.18			vertical
2/	1		Spandril boxed end to eaves 25 mm wrot softwood size 200 x 270 mm overall	16.4.1.6 Filling projecting eaves at gable end
	29.45		Prime only wood general surfaces n..e. 300 mm girth before fixing	29.1.1.2.6
	29.85			
2/	0.20			

				Traditional Roof 8
	29.85 0.40		Knot, prime and stop and two undercoats and one gloss coat to wood general surfaces exc. 300 mm girth externally.	29.1.2.2 Fascias and soffits
			33 000 – 4 500 28 500	
	28.50		Proprietary continuous eaves ventilator fixed to softwood	16.4.1.1
			Gutter Fascia 29 450 3/2/100 600 30 050	
	7.35 3.95 2.60 3.95		Insulation 100 mm thick laid between joists at 400 mm centres, horizontally	33.3.1.3
	30.05		Rainwater gutter, straight half round 100 mm uPVC with combined fascia brackets and clips screwed to softwood	33.5.1.1 Measured over fittings
2/	1		Ancillaries, stop end	33.6.1.1
4/	1		Ditto angles	
4/	1		Ditto outlets	

				Traditional Roof 9
4/	6.00		Rainwater pipes, straight 68 mm diameter uPVC with push fit socket joints and ear piece brackets at 2 m centres plugged to masonry.	33.1.1.1 Pipe length taken over fittings; length assumed
4/2/	1		Extra over for fittings > 65 mm diameter, 2 ends, 68 mm diameter offset bend, 250 mm projection	33.3.1.4
4/	1		Copper wire balloon grating to 65 mm diameter pipe	33.6.1.1

Example 7
Trussed Rafter Roof

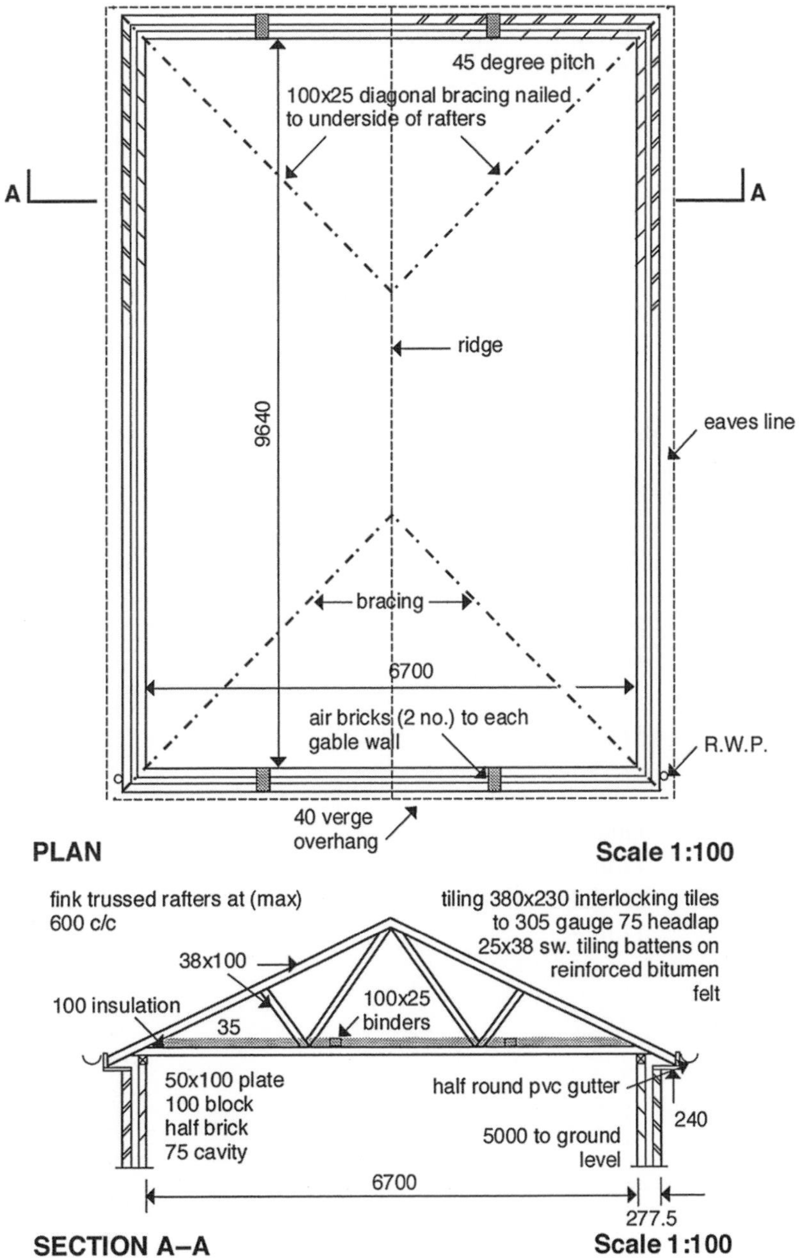
45 degree pitch
100x25 diagonal bracing nailed to underside of rafters
A
A
ridge
9640
eaves line
bracing
6700
air bricks (2 no.) to each gable wall
R.W.P.
40 verge overhang
PLAN
Scale 1:100
fink trussed rafters at (max) 600 c/c
tiling 380x230 interlocking tiles to 305 gauge 75 headlap 25x38 sw. tiling battens on reinforced bitumen felt
38x100
100 insulation
35
100x25 binders
50x100 plate
100 block
half brick
75 cavity
half round pvc gutter
240
5000 to ground level
6700
277.5
SECTION A–A
Scale 1:100

Fig. 30

Example 7

Taking-off list	**NRM2 reference**
Structure	
Wall plates	16.1.1.3
Trusses	16.2.1.2
Binders	16.1.1.1
Bracing	16.1.1.1
Joist anchors	16.6.1.7
Insulation	33.3.1.3
Air bricks	14.25.1.1
Coverings	
Tilings, battens and underlay	18.1.1.1
Ridge	18.3.1.3
Verge tiles	18.3.1.4
Eaves course	18.3.1.2
Tilting fillet	16.3.1.1
Fascia	16.4.1.3
Soffit board	16.4.1.5
Battens	16.3.1.1
Spandril ends	16.4.1.6
Decoration to eaves	29.1.2.2
Eaves ventilator	16.4.1.1
Rainwater goods	
Gutters	33.5.1.1
Fittings	33.6.1.1
Down pipes	33.1.1.1
Fittings	33.3.1.4

				Trussed Roof 1
			Plates 9 640 joint 150 9 790	
2/	9.79		Wall plates 100 x 75 mm treated sawn softwood	16.1.1.3
			trusses 9 640 Clearance 25 ½ rafter 19 – 2/44 _ 8B 600)9 552 16*1 =17 Overhang Eaves 240 Wall 277 517 Less fascia 25 492	
17/	1		Trussed rafters in sawn softwood Fink pattern comprising 38 x 100 mm members, 6700 mm clear span, 492 mm overhang, 35° pitch, jointed with 18 g galvanized steel plate connectors	16.2.1.2 A dimensioned diagram could be included for this item.

				Trussed Roof 2
			<u>Bracing</u> ½ span 6.700 3 350 plate <u>100</u> length on plan 3 450 Roof slope = 3 450 / cos θ = <u>4 212</u> length =/(3.45^2 + 4.212^2) = <u>5 445</u>	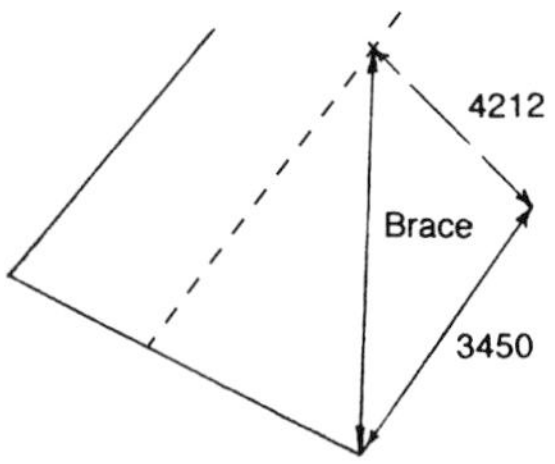
3/ 2/2/	<u>9.64</u> <u>5.45</u>		Roof members pitched 25 x 100 mm treated sawn softwood	16.1.1.1 Binders Bracing
2/2/	<u>1</u>		Galvanised mild steel joist anchor restraint 30 x 5 mm x 300 mm girth, twice bent and twice countersunk drilled	16.6.1.7
			9 640 less trusses 17/38 <u>646</u> 8 994	
	8.99 <u>6.70</u>		100 mm insulation between members at 600mm centres horizontally	33.3.1.3
2/2/	<u>1</u>		Airbricks, square hole pattern, red terracotta size 225 x 150 mm in hollow wall of ½ b facings, 100 mm blks and 75 mm insulated cavity including sealing cavity with slates	14.25.1.1

				Trussed Roof 3
			Covering	
			9 640 Walls 2/277½ 555 10 195 Verge 2/40 80 10 275	
			Wall 277½ 6 700 Eaves 240 Into gutter 50 1 135 ½/ 7 835° = 3 918 ÷ cos 35 Slope length = 4 783	
2/	10.28 4.78		Roof coverings. 35° pitch. 380 x 230 mm Redland interlocking tiles to 75 mm lap, nailed every second course with 48 x 11 gauge aluminium alloy nails. 38 x 25 mm battens galvanised nailed, reinforced bitumen roofing felt to BS747 type IF weighing 15 kg with 75 mm horizontal and 150 mm end laps	18.1.1.1
2/	10.28		Eaves tiling	18.3.1.2
			&	
			Battens 75 x 50 mm extreme, treated softwood, triangular	16.3.1.1
			Verge	
2/2/	4.78		Special tile, including plain tile underclaok	18.3.1.4

				Trussed Roof 4
			Ridge	
	10.28		Ridge, 250 mm diameter half round to match tiling	18.3.1.3
2/	10.20		Fascia board 25 x 150 mm wrot softwood, chamfered and grooved	16.4.1.3
			25 − 13; 240 − 12 = 228	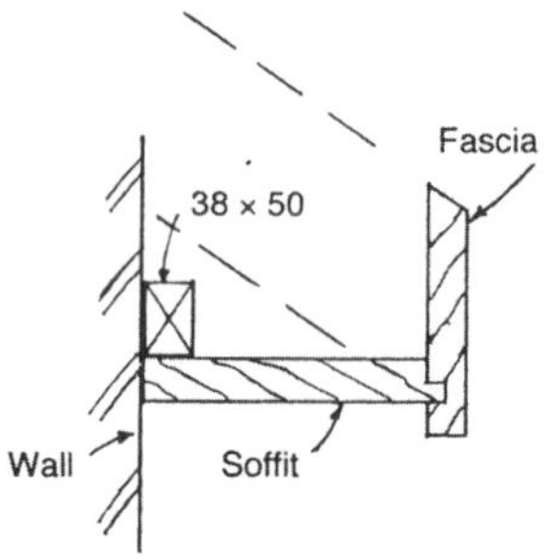
			&	
			Eaves soffit board 25 x 228 mm wrot softwood, rebated	16.4.1.5
			&	
			Eaves ventilator, Redland roofing galvanised steel continuous strip	16.4.1.1
2/	10.20		First fix timbers, battens 38 x 50 mm treated sawn softwood to brick	16.3.1.1
2/2/	1		Spandril boxed end to eaves 25 mm wrot softwood size 240 x 300 mm overall	16.4.1.6
2/2/	10.20		Prime only general surfaces of wood n.e. 300 mm wide, before fixing	29.1.1.2.6 Priming only back of fascia before fixing
2/	0.23			

				Trussed Roof 5
			10 195 Ends 2/300 600 10 795 250 150 25 425	
2/	10.80 0.43		Knot, prime and stop, two undercoats and one gloss coat to wood general surfaces exceeding 300 mm, externally	29.1.2.2 Taken as superficial work to both surfaces as same treatment
2/	10.28		Rainwater gutters, straight 100 mm half round uPVC with combined fascia brackets and clips, screwed to softwood	33.5.1.1 Measured over fittings
2/2/	1		Ancillaries, stop end & Ditto for outlet & Copper wire balloon grating to 68 mm diameter outlet	33.6.1.1
2/	5.20		Rainwater pipes, straight 68 mm diameter uPVC with push fit socket joints and earpiece brackets at 2 m centres plugged to brickwork	33.1.1.1
2/2/	1		Extra over for fittings > 65 mm diameter, two ends, 68 mm diameter offset bend, 250 mm projection	33.3.1.4

Example 8

Asphalt Covered Flat Roof

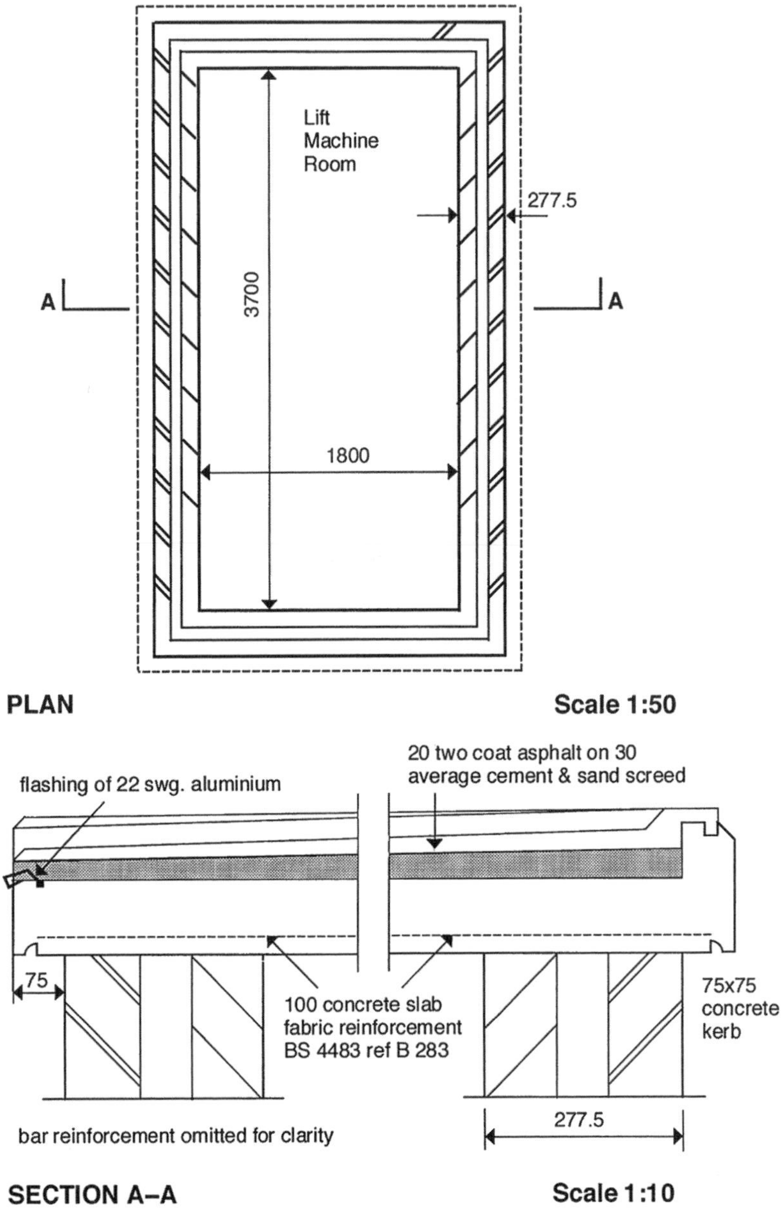

Fig. 31

Example 8

Taking-off list	NRM2 reference
Concrete slab	11.2.1.2
Concrete upstand	11.2.1.2
Soffit formwork	11.15.1.3
Fair finish	
Edge formwork, including groove	11.14.1
Upstand formwork	11.19.1
Surface treatment to upstand	11.8.1
Reinforcement	11.37.1
Asphalt roofing	19.1.1
Concrete surface treatment	11.8.1
Screed	28.1.2.1
Asphalt upstand	19.3.1.4
Flashing	19.1.2
Fixing bearer	16.3.1.1

				Flat roof 1
			Size of slab 1 800 3 700 2/277 ½ 555 555 2/75 150 150 2 505 4 405	
	2.51 4.41 0.10		Reinforced in-situ concrete (25 N-20 mm aggregate) horizontal work, slab n.e. 300 mm thick in structures	11.2.1.2
	9.27 0.08 0.08		4 405 2/2 505 5 010 9 415 less 2/75 150 centre of upstand 9 265	Upstand goes around two short sides and one long side
			Formwork 2 505 4 405 2/6 910 external girth = 13 820 – 4/2/ ½ /75 = 300 13 520	
	3.70 1.80 13.52 0.08		Formwork, fair finish soffits of horizontal work n.e. 300 mm thick, horizontal, height 9 - 10.5 m.	11.15.1.3 Support height is given in stages of 1.5 m. It is assumed that this slab is above a lift shaft. Throating is deemed included.

				Flat Roof 2
	13.82		Formwork, fair finish, to edge of suspended slab, vertical n.e. 500 mm high	11.14.1 Taken to outer edge of upstand total height not less than 250 mm
			Upstand	
			External face a .b. 9 415 – 2/2/75 300 inner face 9 115	
	9.12 0.08		Formwork, to side of upstand beam, rectangular shape 75 x 75 mm	11.19.1 Rebate in top deemed included
	9.12 0.03		Trowelling top surface of upstand	11.8.1
			Reinforcement	
			2 505 4 405 – cover 70 – 2/35 70 2 435 4 335	
	2.44 4.34		Reinforcement, welded fabric ref 8283 weighing 3.73 kg/m^2 with 200 mm laps	11.37.1

				Flat Roof 3
			<u>Coverings</u>	
			<u>Mastic asphalt limestone aggregate to BS 6925 in two coats laid breaking joints and 20 mm thick</u>	
			2 505 4 405 − cover <u>75</u> − 2/75 <u>150</u> <u>2 430</u> <u>4 255</u>	
	2.43 <u>4.26</u>		Coverings, asphalt roofing exceeding 500 mm wide to level and to falls to screeded base	19.1.1
			&	
			Trowel surface of concrete for screed	11.8.1
			&	
			Cement and sand (1:4) screed to roofs, level and to falls n.e. 15^0 from horizontal, 30 mm average thickness in one coat to concrete	28.1.2.1
	<u>9.12</u>		Asphalt skirting to kerb n.e. 150 mm high raking	19.3.1.4 Rounded edge is deemed included.

				Flat Roof 4
			3 700 2/277 ½ 555 4 255	
	4.26		22 swg flashing in aluminium n.e. 100 mm girth, horizontal, fixed with galvanized nails to hardwood	17.5.1
			&	
			Treated hardwood individual support 40 x 25 mm dovetail section and setting in unset concrete	16.3.1.1

Chapter 12
Internal Finishes

Schedules

A schedule of internal finishes listing floor by floor and room by room the ceiling, wall and floor finishes and decorations is an invaluable aid to the measurement of this section of the work. Details of cornices, skirtings, dados and the like can be added to the schedule as required. The schedule brings to light any missing information, enables rooms with similar finishes to be grouped together more easily and reduces the need to refer to the specification during the taking off. Looping through the entries on the schedule as the items are measured ensures that none are missed. At a later stage, the schedule provides a useful quick reference indicating what has been taken without one having to look through the dimensions. An example of a typical schedule is given in Fig. 32. This is an information schedule and may well be provided by the architect. Another schedule could be utilised to measure the finishes. Here, girths and heights of rooms would be recorded and summarised to reduce any repetitive dimensions.

Subdivision

If the ceiling heights and construction vary throughout the building, it may be advisable to keep the measurements on each floor separate. If, however, the room layout and finishes repeat from floor to floor, it is probably more expedient to use timesing. As this will quite possibly be the largest section of taking-off, care should be taken to signpost the dimensions to enable

Willis's Elements of Quantity Surveying, Twelfth Edition. Sandra Lee, William Trench and Andrew Willis.

Published 2014 by John Wiley & Sons, Ltd.

measurements to be traced easily at a later stage. The principal subdivisions in the order recommended for measuring are:

- Floor finishes and screeds (unless taken with floor construction)
- Ceiling finishes (including attached beams with the same finish)
- Isolated beams
- Wall finishes (including attached columns with the same finish)
- Isolated columns
- Cornices
- Dados
- Skirtings.

Note: It may be convenient to measure floors with ceilings if the areas are the same. Confusion may arise, however, if the specification changes are not consistent between the two, and also if deductions have to be made from floors only for such items as stair openings, hearths and fittings.

INTERNAL FINISHES SCHEDULE					
	CEILING		WALLS		
LOCATION	FINISH	DECORATION	FINISH	DECORATION	DADO
Bathroom Kitchen etc.	9.5 mm plasterboard 5 mm skim	2 ct emulsion	13 mm two ct plaster	2 ct emulsion	Glazed tiling 1 m high
			FLOOR		
LOCATION	CORNICE	SKIRTING	BED	FINISH	REMARKS
Bathroom Kitchen etc.	250 mm gth moulded	25 × 100 mm rounded s/w Kps & 3	35 mm cement and sand	Carpet	Area n.e. 4 m^2

Fig. 32

Generally

The measurements for in-situ finishes are taken as those of the base to which the finish is applied. Tiling, on the other hand, is measured on the exposed face. In practice, this makes little difference, as normally the structural room sizes are taken for the measurement of finishes and decoration. One should be careful, however, when measuring tiling to columns as the face area of the tiling will be greater than the structural face area. Generally, no deduction is made for voids within the area of the work when they do not exceed 1.0 m^2.

Floor finishes

In-situ sheet and tile finishings all have to be described as follows:

- Level or to falls less than 15°
- To falls and crossfalls less than 15°
- To slopes more than 15°.

For in-situ sheet and flexible tile floorings, the area measured is that in contact with the base. For rigid tile flooring, the finished face is measured, although in practice, as far as floors are concerned, this usually makes little difference. If the work is laid in bays, this has to be stated, giving their average size. There is no longer a need to keep work in staircase areas and plant rooms separate.

Skirtings are usually measured with internal finishes, as their length relates to that of the walls. Floor coverings in door openings are taken either with doors, as the surveyor dealing with these is familiar with the dimensions, or with the finishes. It should be remembered that dividing strips may be required where the finish changes. Doors are usually hung flush with the face of the wall of the room into which they open; therefore, the floor finish in the opening should be that of the room that shows when the door is closed. Mat frames and mat wells should be measured with floor finishes.

Screeds, if required, are measured with floor finishes as their areas are usually the same. Care should be taken to ascertain the thickness of screeds where the floor finishes vary. Usually, it is necessary to have floor surfaces level; differing finishes may have varying thicknesses, requiring the screed to take up the difference. Sometimes this difference is allowed for in the floor construction.

Timber boarding, plywood or chipboard to floors is measured superficial unless it does not exceed 600 mm wide when it is measured linear. No deduction is made for voids not exceeding 1.0 m^2 within the area of the flooring. Remember that nosings and margins may be required around openings in the floor.

Ceiling finishes

Ceiling areas are taken from wall to wall in each room using the figured structural dimensions. The dimensions are usually best taken over the beams, and the adjustments made later. Care should be taken as the work to the soffit of beams may well be in a different height classification from the main ceiling of a room. If the majority of ceiling finishings are identical, it may be advantageous to measure the ceiling over all the internal walls and partitions. From this measurement will be deducted the total lengths of the internal walls and partitions, possibly obtained from previous dimensions, squared by their respective thicknesses. Treating the surface of concrete by mechanical means, or as it is traditionally known *hacking concrete* to receive plaster, is measured superficially and is conveniently *anded on* to the finishes dimensions. The application of bonding agents or other preparatory work is described with the plasterwork. Plasterboard linings to ceilings are measured in square metres.

Wall finishes

In the case of wall finishings, it is best to schedule or total in waste calculations the perimeters of all rooms having the same height and finish. The areas of the walls may thus be contained in two or three dimensions instead of the large number there would be if each room were measured separately. As in the case of the measurement of walls, openings are generally ignored, the wall finish being measured across. Deductions for openings together with work to reveals would be taken later with the measurement of windows, doors and blank openings. If large openings or infill panels occur from floor to ceiling height, it is sometimes preferable to measure the finishes net. Care must be taken to inform other measurers which openings have been measured net so as to avoid the possibility of a double deduction. As a general rule, if the walling has been measured across an opening, so should the finishes be.

If in-situ wall finishes are applied to brick or block walls, then the raking out of the joints to form a key is deemed included. Preparation of concrete walls is dealt with as mentioned for ceilings in this chapter.

Angle beads and so on

All ends and angles either formed or proprietary are deemed included. Designed movement beads are measured as linear items. Sizes are given in the description, but working finishes to beads are deemed included.

Decoration

The decoration to ceilings and walls should be measured with the finish. If the decoration is all the same this can be anded on to the plaster or plasterboard description. If, however, there is a mixture of several types of decoration, such as emulsion, gloss paint and wallpaper, it will probably be best to subdivide the plaster or plasterboard measurement so that the appropriate decoration can be anded on. Where all is, say, emulsion except wallpaper in one or two rooms, it is usually more satisfactory to measure the whole as emulsion in the first instance and then to deduct the emulsion and add wallpaper for the appropriate area. Isolated paintwork where the girth does not exceed 300 mm or where the area does not exceed $1.0\,m^2$ is measured linear or enumerated, respectively. Paintwork to walls, ceilings, beams and columns is described as work to general surfaces and, if the same, can be added together in the bill. Work to ceilings and beams exceeding 3.5 m above floor level has, however, to be so described in stages of 3 m. Papering has to be described in the two categories of ceilings with beams and walls with columns, the same rule regarding high ceilings applying.

Cornices and coves

The measurement of cornices and coves is usually straightforward, as they normally run all round the room without interruption. Both in-situ and prefabricated types are measured the length in contact with the base, using the length already calculated for the perimeter of the walls. The measurement must include returns to projections and to beams if these occur in the ceiling. Ends, angles and intersections are deemed included. Paintwork to cornices does not have to be kept separate if it is the same as that to the walls or ceilings. In fact, it is questionable whether any adjustment should be made to the wall and ceiling decoration if this is the same, particularly if the cornice is contoured.

Skirtings

There is a difference of opinion as to whether skirtings should be measured net or measured gross across openings. The disadvantage with measuring net is that the taker-off measuring finishes will have to ascertain the width of openings and possibly the depth of reveals. On the other hand, wall decoration, as opposed to plaster, should be deducted behind the skirting. Thus, if the skirtings are measured gross with the deduction of decoration behind, when the doors are measured a deduction of plaster and decoration is usually made for the height of the opening. This results in the decoration being deducted twice behind the skirting, and therefore an addition of decoration will have to be made for the length by the height of the skirting across the door opening.

Timber skirtings are measured linear, usually using structural wall face dimensions, but care will have to be exercised with measurements to isolated piers and the like, as the length will be increased. The size and shape of the skirting and method of fixing if specified are given in the description, but mitres, ends and intersections are deemed included. If the skirting is fixed to grounds, no deduction of plaster is made for grounds. Dado and picture rails are measured in the same way as skirtings.

In-situ and tile skirtings are measured as linear items, stating the height and thickness in the description. Fair and rounded edges are deemed to be included, as are ends and angles.

Wall tiling

Firstly, it is necessary to ascertain whether the wall tiling is fixed with adhesive or bedded in mortar. The specification may, particularly in the case of a tiled dado, call for the wall behind the tiling to be plastered like the remainder of the wall and for the tiles to be fixed to the plaster with adhesive. Alternatively, there may be a cement and sand screed behind the tiling, with the tiles either bedded in mortar or fixed with adhesive. If there is a mixture of wall tiling and plaster finish on a wall, it is usually necessary for the back of the tiles to be flush with the face of the plaster. If special tiles with rounded edges are specified at top edges and external angles these are measured as linear items extra over the main work. The rules for measurement of tiling follow closely those for in-situ finishes, but attention is drawn to the 'Generally' section of this chapter relating to measurements.

Internal partitions

The measurement of timber stud partitions is described in Chapter 9. Metal stud partitions can be measured in a similar way.

Dry wall linings

Plasterboard dry linings are measured square, stating if the width on face is over or under 300 mm wide. Internal and external angles are measured as linear items.

Example 9

Internal Finishes

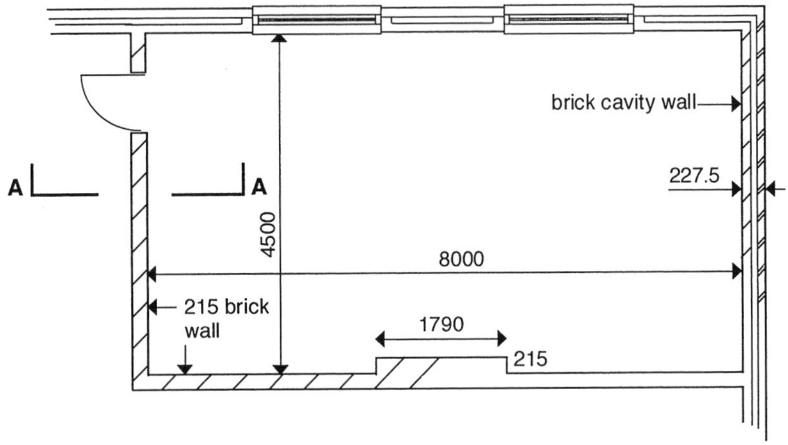

PLAN **Scale 1:100**

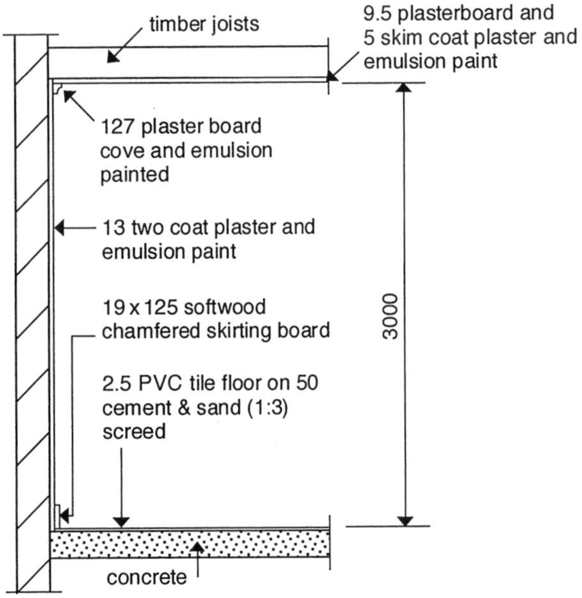

SECTION A–A **Scale 1:50**

Fig. 33

Example 9

Taking-off list	**NRM2 reference**
Ceilings: plasterboard and plaster	28.9.2
emulsion	29.1.2.1
Floor: floor tiles	28.2.2
screed	29.1.1
Wall plaster	28.7.2
Plaster to pier	28.7.1
Emulsion	29.1.2.1
Plaster cove	28.17.1
Skirting	22.1.1
Paint to skirting	29.1.1.1
Adjust paint for skirting	

				Internal Finishes 1
	8.00 4.50		Finishes to ceilings, width exceeding 600 mm, 9.5 mm thick gypsum plasterboard fixed with rustproofed nails fixed to softwood, 5 mm thick coat thistleboard finish plaster &	28.9.2
			Two coats emulsion paint to general surfaces over 300 mm wide, internal &	29.1.2.1
			Floor finish, PVC tile 225 x 225 x 2.5 mm, over 600 mm wide &	28.2.2
			Screed, cement and sand to floors, over 600 mm wide, level and to falls not exceeding 15° from horizontal	29.1.1
	1.79 0.22		[Pier] Deduct Last four items	Deduction for area displaced by the pier. Note that the minimum deduction rule does not apply as the void is at the perimeter.

				Internal Finishes 2
			8 000 4 500 2/12 500 = 25 000 2/215 430 25 430	
	25.00 3.00		Finish to walls exceeding 600 mm wide, plaster in two coats 13 mm thick to brick base.	28.7.1
2/	3.00		Ditto not exceeding 600 mm wide	28.7.2
	25.43 3.00		Two coats emulsion paint as before	29.1.2.1
			Coves	Angle beads deemed included
	25.43		Gyproc coving 127 mm girth fixed with adhesive	28.1.7.1 Angles deemed included

				Internal Finishes 3
	25.43		Wrot softwood skirting 19 x 125 mm chamfered, plugged and screwed to brickwork	22.1.1
			&	
			Knot, prime and stop and three coats gloss paint on general surfaces woodwork, less than 300 mm girth, internal	29.1.1.1
			&	
			Prime only general surfaces of woodwork, n.e. 300 mm girth before fixing	29.1.2.1.6
			&	
			Deduct Two coats emulsion paint to plaster surfaces as before $\times$ 0.13 = ______ m^2	

Chapter 13
Windows and Doors

Subdivision

The measurement of windows and doors can be conveniently subdivided as follows:

- Windows
- External doors
- Internal doors
- Blank openings.

Whilst it is not essential to take external and internal doors separately, it will probably be found that they are different types of doors with different finishes, and therefore they may have little in common.

Windows
- Timber, metal or unplasticised polyvinyl chloride (uPVC) casement and fixing
- Glass (if not included in the above)
- Ironmongery (ditto)
- Decoration (if required).

Opening adjustments
- Deduction of brickwork and blockwork and external and internal finishes
- Support to the work above the window
- Damp proofing and finishes to the head externally and internally
- Ditto to the external and internal reveals
- Ditto to the external and internal cills.

Willis's Elements of Quantity Surveying, Twelfth Edition. Sandra Lee, William Trench and Andrew Willis.

Published 2014 by John Wiley & Sons, Ltd.

External doors

- The same items would be measured for external doors as shown above for windows.

Internal doors and blank openings

- Again, this will be similar as given for Windows but in place of cills, flooring in the opening will be required. Damp proofing around the opening will not be necessary.

Steps to external doors may be measured at this stage along with porches, canopies or other special features.

If these items are followed through systematically in each group there will be less chance of items being missed.

Schedules

As with internal finishes measurement and for similar reasons, it is usually prudent to prepare a schedule of windows and doors before commencing measurement. Before preparing the schedule, the windows should be lettered or numbered on the floor plans and the elevations. If there are several floors to the building, then a system of numbering should be devised which enables windows on a particular floor to be located readily. For example, a letter could be allocated to each floor followed by a number for each window, the numbering starting at a particular point on each floor and proceeding in, say, a clockwise direction round the building. Whilst windows are usually scheduled floor by floor, there may be good reason to work by elevations or by window types or even by a combination of all three. The total number of windows entered on the schedule should be checked carefully with the total number shown on the drawings to ensure that none is missed. A further check must be made to ensure that the windows shown on the elevations tie up with those shown on the plans. Discrepancies may occur because clerestory windows or windows to mezzanine floors are sometimes shown on the elevations but not on the plans. Any differences found should be mentioned to the architect and the matter resolved before commencing measurement. The schedule should aim to set out the details of each window so that it is hardly necessary to refer to the drawings during measurement. Usually it should be possible to measure together in one group all windows of one type irrespective of their size, the wall thickness and so on.

A note should be made at the commencement of the measurement for each group of the numbers of windows being dealt with; care should be taken throughout that this total is accounted for in each item. A common fault of beginners is to separate the entire measurement of windows of the same type but of different sizes or in different thicknesses of walls. Such a method may give less trouble but will probably take longer. Although the grouping of windows of different sizes may require more concentration and care, proper use of the schedule will considerably simplify the work.

Doors on the plans should be lettered or numbered socially as described for windows above. External doors may be included with the window numbering, and internal blank openings included with internal doors.

Timesing

Window dimensions will probably contain a fair amount of timesing, and great care is necessary to ensure that this is done correctly. As each of the subdivisions of measurement is completed, it is advisable to total the timesing of each item to ensure that it equals the number of windows being measured. If, for instance, 20 windows are being measured in a group, timesing of deductions, lintels, cills, and the like should total 20 unless a change in specification or design for a particular window requires a smaller number.

Special features

Usually any special features that definitely relate to the window (such as small canopies above or decorative brickwork underneath) will be measured with the window.

Dormer windows

In the case of dormer windows, the window itself may be taken with the other windows whilst the adjustment of the roof would normally be taken with the roofs. This division will generally be found to be convenient, particularly if the roofs and windows are being measured by different persons. The opening for a dormer window may sometimes be partly in the wall and partly in the roof, in which case the wall adjustment would be made with the window measurement, in the same way as for the other window openings.

Adjustments

When measuring a group of items, the advantage of taking initially the same description for similar work and then making an adjustment for small differences often becomes evident in the measurement of windows.

For example, if all the windows are glazed in clear glass except for a small proportion which have patterned glass, the simplest way of measuring may be to take all as clear glass and then, checking over carefully with the schedule, make adjustment for those that need patterned glass. If, as is not impossible, the measurement of the patterned glass to one or more windows should be missed, then if clear glass has been measured to them all, the error will be much less than if none had been measured. Similarly, when making adjustments for the openings, different decoration may be applied to the walls. If the predominating one is chosen for the deductions to all windows then, as an adjustment, the true decoration may be deducted where appropriate and the predominating finish, deducted earlier, added back. Apart from minimising the effect of errors, this method also facilitates the grouping of descriptions under the same measurements.

Windows and doors

Timber, metal and uPVC windows and frames or doors and frames are enumerated and described with a dimensioned diagram. A reference to a catalogue or standard specification may remove the need to provide a diagram. The method of fixing must be shown unless this is at the discretion of the contractor. Timber window boards and cover fillets are measured as linear items, stating their cross-section dimensions and labours in the description.

Glass

The size of each pane of glass can sometimes be obtained from manufacturers' catalogues; otherwise it has to be calculated from the overall size of the window, deducting for frames, mullions and so on. Raking and curved cutting to glass is deemed to be included Adjacent panes of glass required to align with each other have to be so described. One of the rare occasions when waste is allowed for in measurement occurs in glazing: panes of irregular shape are measured as the smallest rectangle from which the pane can be obtained. Bedding the edges of glass in strips or channels is included in the description.

If the glazing is hermetically sealed double glazing units or special glasses are specified, then the panes are enumerated, stating their size and the airspace width.

Ironmongery

Each item of ironmongery, if not supplied with the window, is enumerated and described giving the nature of the background to which it is fixed. The description of ironmongery is simplified by reference to manufacturers' catalogues, being careful to give sizes, material and finish if alternatives are listed. Frequently it is convenient to include a PC sum for the supply of ironmongery and measure out the fixing. This avoids the necessity to give a full description for the ironmongery, which is impossible at the measurement stage if a selection has not been made.

Decoration

Decoration is measured as a superficial item, the measurement being taken over frames, mullions, transoms, cills and glass. The description would be for decoration to glazed surfaces irrespective of pane sizes.
The measurement is deemed to include such items as paint on opening edges and the consequential extra frame, cutting in and work on glazing beads. Decoration that is external has to be so described. Priming only to backs of frames before fixing is measured as a linear item if not exceeding 300 mm girth.

Openings

Unless the opening has rebated reveals, the same dimensions can be used for the deduction of the wall, cavity and external and internal finishes. After making this adjustment, it is best to consider the perimeter of the opening in the order of head, jambs and cill. Precast concrete lintels are enumerated; the size, shape and reinforcement is included in the description. In-situ concrete lintels are measured as isolated beams; the cubic measurement of the concrete and the measurement of the formwork and reinforcement each have to be taken separately. Brickwork displaced by lintels is only deducted for height to the extent of full courses displaced and for depth into the wall to the extent of full half-brick beds displaced. Damp-proof courses to heads forming cavity trays are so described. Proprietary steel lintels are enumerated, giving the manufacturer's reference. Facework to arches is measured as a linear item, stating the number. Plaster to the reveal, if not exceeding 300 mm wide, is measured as a linear item. Decoration to the reveal is measured as the work to the adjacent wall. At the jambs, cavity closing is measured as a linear item; facework to the reveals is deemed included but vertical damp proof courses are measured. Precast concrete cills are measured in the same way as lintels with the same rules for adjustments. Facework to cills is measured as a linear item, stating whether it is set weathering and the dimensions.

Example 10

Window

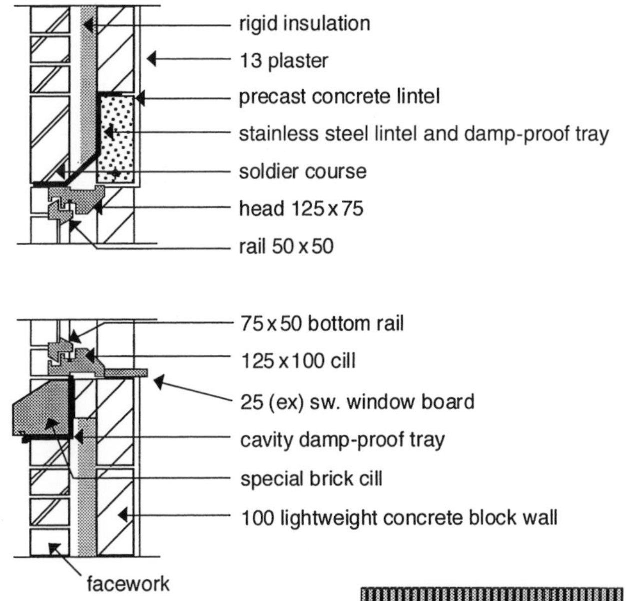

SECTION
Scale 1:20

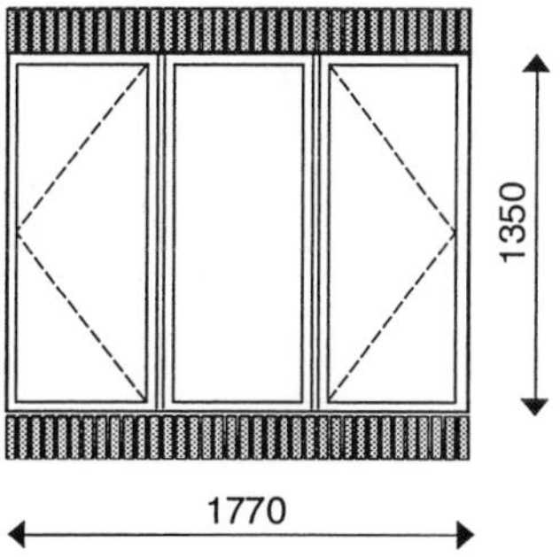

ELEVATION n.t.s.

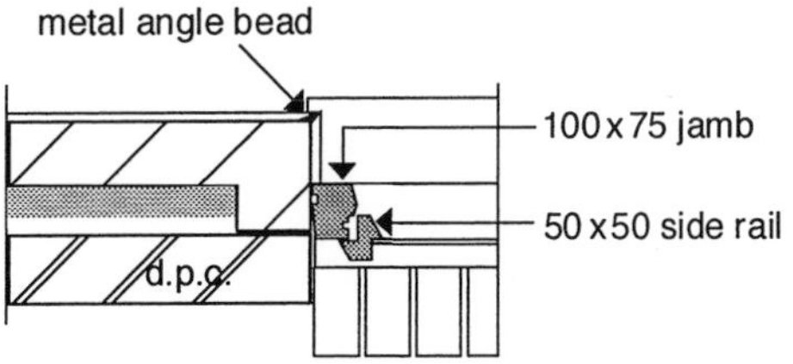

PLAN
Scale 1:20

Fig. 34

Example 10

Taking-off list		NRM2 reference
Softwood window		23.1.1
Bed frame		Deemed included
Window board		22.5.1
Glazing		23.8.1
Paint to windows		29.2.2, 29.1.1
Paint to window board		29.1.1.1
Adjust wall construction:	facings	
	block	
	cavity	
	insulation	
	plaster	
	decoration	
Lintel		14.25.1
Cavity tray		14.18.1.3
Brick arch		14.6.1
Adjust brick for arch		
Plaster to reveals		28.7.1
Angle beads		Deemed included
Paint to reveals		29.1.1.1
Close cavity		Only if proprietary
Damp proof course		14.16.1.1
Cill		14.12.1.1
Adjust brick for cill		
Damp proof course		14.16.1.3

				Windows 1
	1		Softwood window size 1 770 x 1 350 mm (see Figure 33), the frame fixed with four galvanised mild steel fixing cramps screwed to back of frame and built into brickwork, with easy clean hinges, standard brass casement stay and two pins and brass casement fasteners	23.1.1 Bedding frames deemed included
	1.77		Wrot softwood window board 25 x 150 mm rebated and rounded one edge, screwed and pelleted to masonry	22.5.1
			Glass Width 1 770 Less Frame 2/40 = 80 Mullions 2/30 = 60 — 140 1 630 Fixed width ÷ 3 = 543 Less Side rails 2/20 — 40 Open width = 503	
			Height 1 350 Less Head 40 Cill 50 — 90 Fixed 1 260 Less Top 20 Bottom 30 — 50 Open light 1 210	Deduction for area displaced by the pier. Note that the minimum deduction rule does not apply as the void is at the perimeter.

				Windows 2
	2		Sealed double-glazed units in clear glass size 500 x 1 210 mm as manufacturer's specification and glazing with springs and putty to wood	23.8.1
	1		Ditto size 540 x 1 260 mm	
2/	1.77 1.35		Knot and prime to glazed wood surfaces irrespective of pane size, over 300 mm girth off site prior to fixing	29.2.2.7
	1.77 1.35		Two undercoats and one top coat gloss paint to glazed surfaces irrespective of pane size, over 300 mm girth, internal & Ditto external	29.2.2.1

				Windows 3
	1		Knot and prime only general surfaces woodwork, isolated areas not exceeding 1 m^2, on site prior to fixing	29.1.3.1
			&	
			Two undercoats and one top coat gloss paint to general surfaces woodwork, isolated areas not exceeding 1 m^2, internal	29.1.3.1.6
			<u>Opening Adjustment</u>	
	1.77 <u>1.35</u>		<u>Deduct</u>	
			Wall in facings as before	Descriptions for the deductions need only be sufficient to identify the items measured previously.
			&	
			<u>Deduct</u>	
			Forming cavity in hollow wall 75 mm wide as before	
			&	
			<u>Deduct</u>	
			Wall insulation as before	
			&	
			<u>Deduct</u>	
			100 mm thick block wall as before	

				Windows 4
	1.77 1.35		Deduct 13 mm plaster to walls as before & Deduct Two coats emulsion paint to plaster surfaces as before	
			Head Lintel 1 770 Bearing 2/150 300 2 070	
	1		Precast concrete (27 N/mm^2 - 20 mm agg.) lintel rectangular section 2 070 x 100 x 210 mm, reinforced with one 16 mm diameter mild steel bar and build into blockwork in gauged mortar (1:1:6) & 3 x 350 mm girth stainless steel combined lintel and cavity tray 2 070 mm long and build into blockwork in gauged mortar (1:1:6)	14.25.1

				Windows 5
	1.77		Flat soldier arch 215 mm wide on face in facing bricks in gauged mortar (1:1:6) half brick thick, width of exposed soffit 20 mm, including pointing (in Nr 1)	14.6.1
	2.07 0.21		Deduct Block wall 100 mm thick as before & Deduct Plaster 13 mm thick to walls over 600 mm girth as before	Deduction of one complete course of blocks Assumed different specification for plaster to concrete
	2.07 1.77		Plaster to walls n.e. 600 mm wide, two coats 13 mm thick to concrete, including bonding agent	28.7.1
	1.77 0.10		Two coats emulsion paint to plastered walls as before	
	1.77 0.23		Deduct Walls in facings brick, half brick thick as before	

				Windows 6
			<u>Jambs</u>	
2/	1.35 <u>0.08</u>		100 mm thick block wall as before	Extra blockwork measured for closing cavity
2/	<u>1.35</u>		Damp-proof course < 300 mm wide, vertical, pitch polymer bedded in gauged mortar (1:1:6)	14.16.1
			&	
			Plaster to walls n.e. 600 mm wide, two coats 13 mm thick to blockwork	28.7.1
2/	1.35 <u>0.10</u>		Two coats emulsion paint to plastered walls as before	29.1.1.1
			<u>Cill</u>	
	<u>1.77</u>		Facework cill, horizontal, purpose made splayed bricks 65 x 150 x 140 throated and set projecting 50 mm and pointing to exposed faces	
	1.77 <u>0.15</u>		<u>Deduct</u> Walls in facings brick, half brick thick as before	
			1 770 2/150 <u>300</u> <u>2 070</u>	
	2.07 <u>0.25</u>		Damp-proof course exceeding 300 mm wide, horizontal a.b.	14.16.3

Example 11

Internal Door

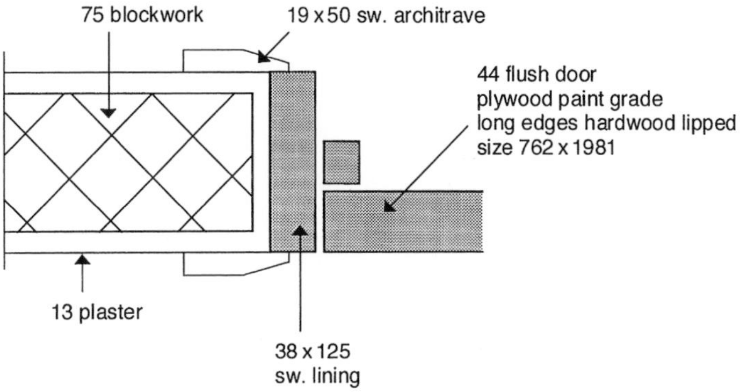

SECTION
Scale 1:10

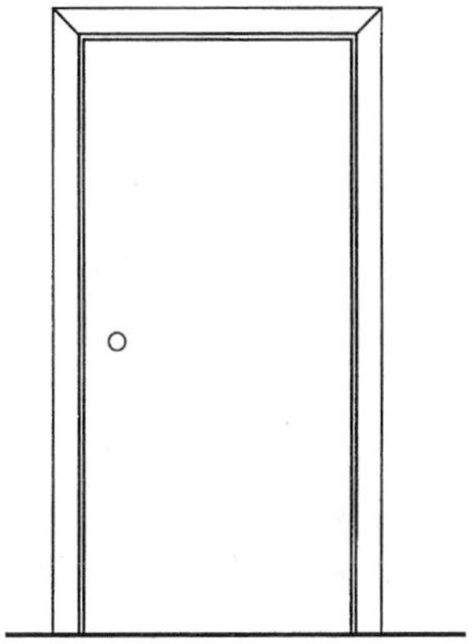

ELEVATION n.t.s.

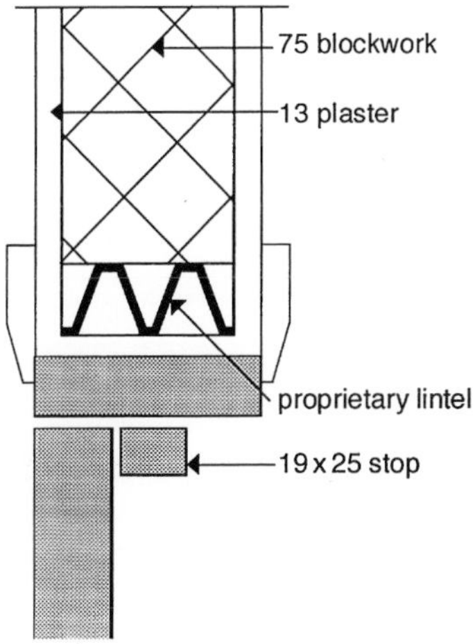

PLAN
Scale 1:10

Fig. 35

Example 11

Taking-off list	NRM2 reference
Door unit	24.2.1
Decoration	29.1.2.1
Ironmongery: supply and fix	22.1.1
Linings	24.10.1
Stops	24.11.1
Architraves	22.2.1
Decoration	29.1.1.6
Adjust wall for door: block plaster decoration	29.1.1
Lintel	14.25.1
Adjust skirting for door	

				Internal Doors 1
	1		40 mm flush door 762 x 1 981 mm with painted grade plywood and hardwood lipped on long edges as diagram (Fig. 33) 762 1 981 40 40 802 2 021	24.2.1
2/	0.80 2.02		Knot, prime and stop, two undercoats and one gloss top coat paint to wood general surfaces exceeding 300 mm wide	29.1.1 The door thickness is added to cover the painting of edges.
			Supply and fix the following ironmongery to hardwood-lipped, softwood flush doors	24.16.1
	1		Pair 75 mm pressed steel butt hinges	

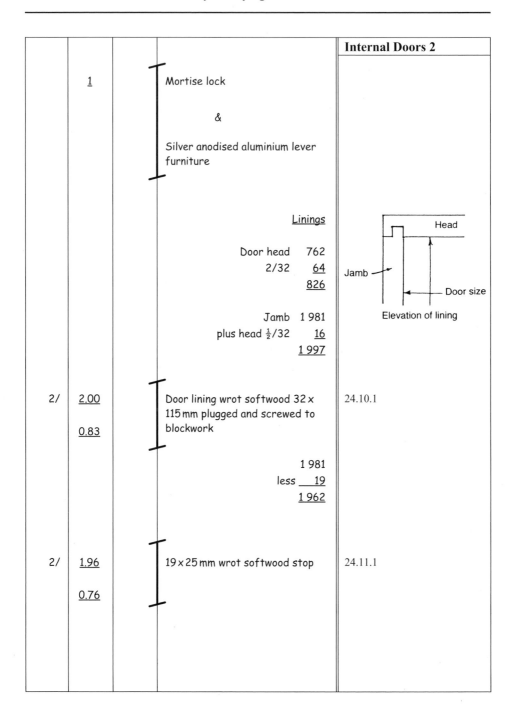

				Internal Doors 2
	1		Mortise lock & Silver anodised aluminium lever furniture	
			Linings Door head 762 2/32 64 826 Jamb 1 981 plus head ½/32 16 1 997	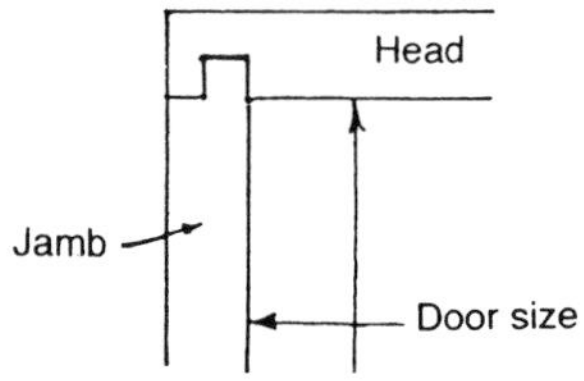 Elevation of lining
2/	2.00 0.83		Door lining wrot softwood 32 x 115 mm plugged and screwed to blockwork	24.10.1
			1 981 less 19 1 962	
2/	1.96 0.76		19 x 25 mm wrot softwood stop	24.11.1

				Internal Doors 3
			<u>Architrave</u>	
			Door head 762	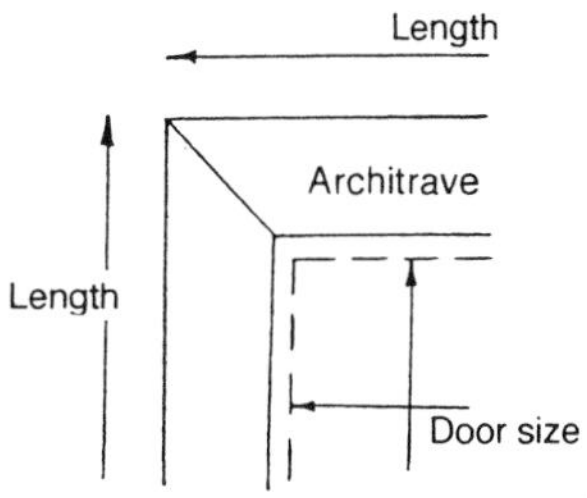
			Width allowance 2/10 20	
			Width of arch 2/75 <u>150</u>	
			932	
			Jamb 1 981	
			plus head 10	
			<u>75</u>	
			2/2 066 <u>4 132</u>	
			<u>5 064</u>	
	<u>5.06</u>		Architrave wrot softwood 25 x 75 mm, splayed and rounded	22.2.1
			<u>Painting</u>	
2/	<u>0.83</u> <u>2.00</u> <u>0.76</u> <u>1.96</u> <u>5.06</u>		Prime only wood general surfaces n.e. 300 mm girth before fixing	29.1.1.6
			<u>Girth</u>	Girth includes architraves, linings and stops.
			Lining width 115	
			Stop depth 2/10 20	
			Lining depth 2/½/38 38	
			Architrave width 2/75 150	
			Architrave depth 2/25 <u>50</u>	
			<u>373</u>	
	4.82 <u>0.37</u>		Knot, prime and stop two undercoats and one gloss coat paint to wood general surfaces exceeding 300 mm girth	29.1.2

				Internal Doors 4
			Opening 762 1 981 2/32 64 32 826 2 013	
			Deduct	
	0.83 2.01		75 mm thick block wall a.b.	14.1.20
2/	0.83 2.01		Deduct Two coats plaster to block wall a.b. & Deduct Two coats emulsion paint general plaster surfaces	28.7.2 29.1.2 The small amount of paint behind the skirting has not been adjusted, although adjustments could be made if the finish were expensive.
			Lintel 826 Bearing 2/75 150 976	
	1		Proprietary coated galvanised steel lintel type ______ 1 m long and building into brickwork	14.25.1 Standard length. No deduction of wall as full course is not displaced.

				Internal Doors 5
2/	0.93		Deduct 19 x 125 mm wrot softwood skirting. & Deduct Knot, prime and stop three coats oil-based paint to wood general surfaces not exceeding 300 mm girth & Deduct Prime only wood general surfaces n.e. 300 mm girth before fixing & Add Two coats emulsion paint general plaster surfaces x 0.13 = ________ m^2	22.1.1 The emulsion paint is deducted twice with skirting and once with door opening adjustment.
			Adjustment for flooring in opening would be measured if not taken with finished take-off.	

Chapter 14
Reinforced Concrete Structures

Generally

This type of work involves the measurement of a number of components comprising a combination of concrete, steel reinforcement and temporary support known as *formwork*.

The measurement of bar reinforcement, although not difficult, is beyond the scope of this book. It is, however, measured linear, and then the linear measurements are converted to a weight prior to billing. The calculation of the number of bars generally follows the approach used in dealing with rafters or floor joists. The distance between the first and last bars is divided by the spacing of the bars, and by adding one to the result the number of bars can be calculated. Fabric reinforcement is measured superficially to the actual area covered, the estimator allowing in the price for the loss of material due to the laps. In practice, the required reinforcement is usually included on an engineer's schedules and therefore just needs to be 'extended'. Total lengths of each bar type are calculated, which can then be direct billed.

The measurement of a reinforced concrete structure requires a clear and logical approach to the order of the take-off so that items are not missed. The work may be subdivided into several parts of the building, and then taken off floor by floor, measuring the concrete, formwork and reinforcement for each floor. In some instances, it may be more practical to measure all of one type of structural component, such as columns, throughout the building. It is also advisable to measure the associated formwork after the concrete so that it is not overlooked, and in any case the same dimensions are usually applicable. Again, a schedule can be of great assistance in simplifying the take-off and reducing any repetitive dimensions.

Willis's Elements of Quantity Surveying, Twelfth Edition. Sandra Lee, William Trench and Andrew Willis.

Published 2014 by John Wiley & Sons, Ltd.

Columns

Columns should be measured between floors; in counting the number of columns, one column is taken at each grid point on every floor level. On a two-storey building, there would therefore be two columns at each point, one on the ground floor and one on the first floor. The concrete is measured in cubic metres. The formwork to simple columns is measured as a superficial item, taking the concrete length multiplied by the girth and stating the total number of columns. This allows the estimator to value the cost of forming the column kickers, which are not measurable, and to add an allowance per square metre to the column formwork cost.

Structural floors and roofs

Structural floors and roofs are measured to the overall dimensions to the outside edge of the columns or beams. The slab concrete is measured as a cubic item, the thickness of the slab being identified as not exceeding 300 mm or as exceeding 300 mm (slabs in the same class can be grouped together).

The soffit formwork is measured as a superficial item to the net area excluding beam soffits and column heads. Further information is given in the description, such as the thickness of the concrete (in the stages of not exceeding 300 mm thick, over 300 but not exceeding 450 mm thick or over 450 mm thick) and the support height (in 1.5 m stages).

Depending on the complexity of the floor plan, a typical bay size could be measured and then be multiplied by the number of bays, or the area could be measured overall and then the beam and column areas deducted. Formwork will also need to be measured to the slab edges and should be measured as a linear item, stating the height in accordance with NRM2, Section 11.14.1.

Beams

Beams should be measured between the columns. Where these are attached to concrete slabs, the volume of concrete is added to the slab concrete. This does not affect the description of the slab because it is the general thickness that is given in the description.

The formwork to the sides and soffit of the beams is measured as a superficial item, using the concrete length and cross-section sizes to obtain the superficial area.

Where beam sides and floor edge coincide, the floor edge is added to the girth of beam formwork. There is no need to make any deduction from the measurements for junctions of members.

Walls

Walls are measured in the same way as slabs, and it should be noted that they include attached columns in the same way that the beams were added to the slabs. It is common practice to cast a small part of the wall with the slab. This is called a *wall kicker*; although it makes no difference to the concrete measure, it does affect the formwork. A linear item is taken for the wall kicker and measured on the centre line of the wall; the price is to include for formwork to both sides. Formwork to single sides of walls needs to be identified and kept separate.

Reinforcement

NRM2 requires that the weight of bars shall include that for bends and hooks: for each hook, one can usually add nine times the diameter of the bar rounded to the next 10 mm. If the length of the concrete is used for calculating the length of the bar, then the concrete cover must be deducted; in other words, in the case of a 12 mm bar, 110 mm must be added at each end less the amount of concrete cover.

Example 12

Concrete Frame

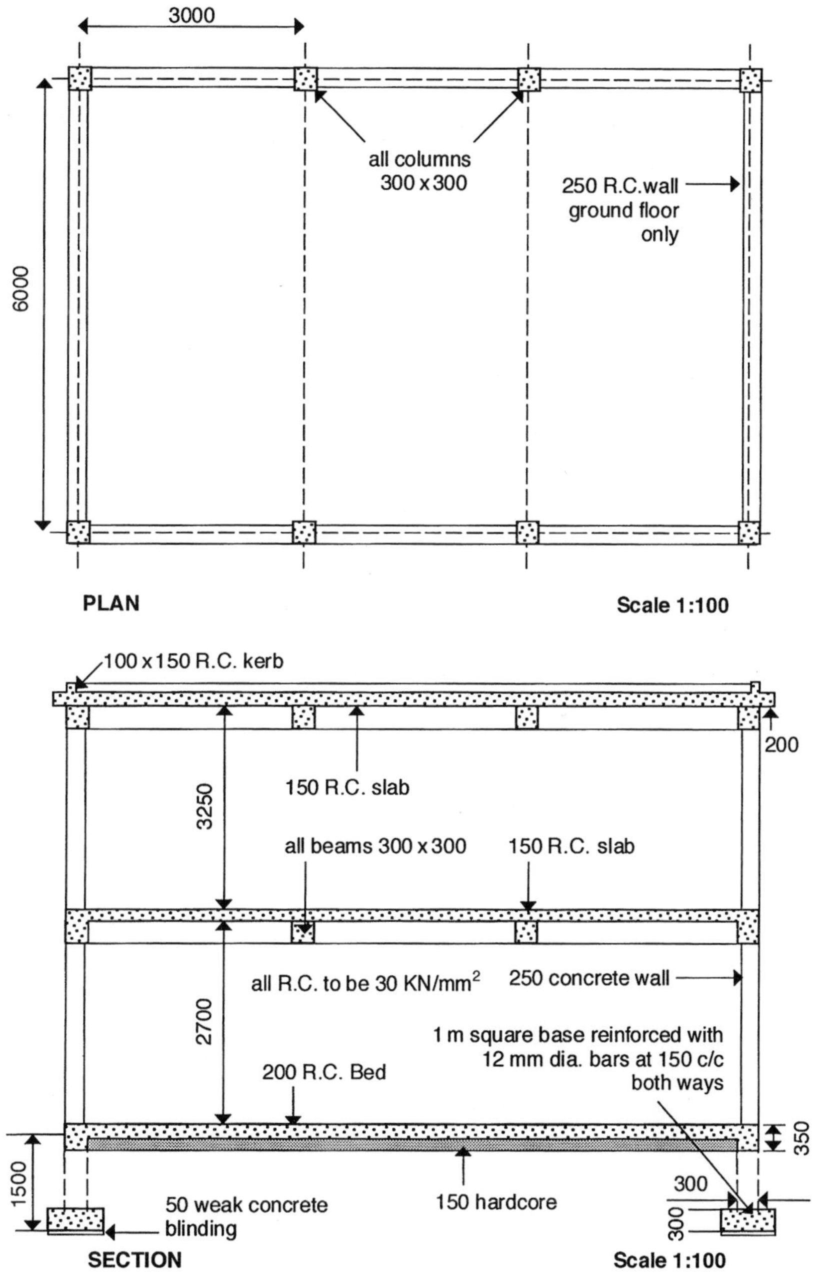

Fig. 36

Example 12

Taking-off list	**NRM2 reference**
Topsoil excavation	5.5.2
Topsoil disposal	5.10.1
Foundation excavation	5.6.2
Disposal of soil	5.9.2
Adjust for backfilling	5.11.2
Earthwork support	Not measurable
Surface treatment	Not measurable
Blinding concrete	11.1.1.1
Concrete bases	11.2.1.2
Reinforcement	11.33.1.1
Concrete columns	11.5.1.1
Formwork to columns	11.20.1
Concrete bed	11.2.1.2
Surface treatment	11.9.1
Hardcore bed	5.12.3
Surface treatment to hardcore	Not measurable
Thickening to bed	11.2.1.2
Formwork to edge of bed	11.14.1
Working space	Not measurable
Concrete wall	11.5.2.1
Attached columns	Include with wall
Formwork to wall	11.22.1
Formwork to attached columns	11.21.1
Wall kicker formwork	11.32.1
Concrete slabs	11.2.1.2
Surface treatment	11.9.1
Attached concrete beams	Include with floor slab
Soffit formwork	11.15.1.1
	11.15.1.2
Beam formwork	11.18.1.1
Formwork to edge of slab	11.18.2.1
Concrete upstand	11.2.1.2
Formwork to upstand	11.19.1

				Concrete Frame 1
			ALL CONCRETE TO BE 30 KN/mm^2	
			Substructure	
			Depth 1 500 blinding 50 1 550 - topsoil 150 1 400	
			W B 6.000 3/3.00 9.000 0.300 ½ col x 2 0.300 6.300 9.300	Calculation of overall dimensions to outer edge of concrete foundation
			Bases 1 000 less column 300 2) 700 spread 350	
	9.30 6.30		Site preparation, remove topsoil average 150 mm thick	5.5.2
8/	1.00 0.35		&	Additional excavation for projection of bases
4/	0.65 0.35		Retaining excavated material on site in temporary spoil heap average 100 m from excavation	5.10.1
			x 0.15 = m^3	Compaction of surfaces is deemed included.

				Concrete Frame 2
8/	1.00 1.00 1.40		Excavation, commenced at stripped ground level, foundations depth n.e. 2 m & Disposal of excavated material off-site	5.6.2 5.9.2 Earthwork support deemed included
8/	1.00 1.00		Weak concrete bed (1:12) n.e. 150 mm thick, poured on or against face of earth x 0.05 = m^3 & Reinforced concrete mix C12, horizontal work ≤ 300 mm thick, foundation poured against face of earth x 0.30 = m^3	11.1.1.1 11.11.2.1.2
8/	0.30 0.30 1.20		Reinforced concrete as before, vertical work in structures ≤ 300 mm thick	11.5.1.1 Height taken from top of pad to underside of bed

				Concrete Frame 3
8/	1.00 1.00 1.05		Filling to excavation exc. 0.25 m thick with excavated material compacted in 150 mm layers & Deduct Disposal excavated material of site	5.11.2 Height of excavation less depth of concrete and blinding
8/	0.30 0.30 1.05		Deduct both last	For stub columns
			Reinforcement Base 1 000 - cover 2/50 100 centres 150) 900 bars = 6 + 1 = 7	Disposal of surface water deemed included.
8/2/ 7/	0.90		Mild steel bar reinforcement 12 mm diameter, straight	11.3.3.1.1 Assumes a mat of straight bars in both directions in each base To be weighed up prior to billing
8/	1.20 1.20		Formwork to square columns, isolated, height n.e. 1.5 m (No. 8)	11.20.1

				Concrete Frame 4
			Overall length 9 300 - edge beams 2/300 600 8 700	
			Overall width 6 300 - edge beams 2/300 600 5 700	
	8.70 5.70		Imported filling beds over 50 n.e. 250 mm thick, hardcore as specified x 0.15 = m^3	5.12.2
	9.30 6.30 0.20		Reinforced concrete 30 KN/mm^2, horizontal work ≤ 300 mm thick bed	11.2.1.2
2/	5.70 0.30 0.15			Thickenings are included with beds. Taken between columns.
2/3/	2.70 0.30 0.15			

				Concrete Frame 5
				Surface packing deemed included.
2/	9.30 6.30		Power float surface of concrete	11.9.1
	9.30 6.30		Formwork to edge of horizontal work ≤ 500 mm high	11.14.1
				Working space allowances deemed included
			Superstructure	
6/	0.30 0.30 2.70		Reinforced concrete 30 KN/mm^2 vertical work in structures ≤ 300 mm thick	11.5.1.1
8/	0.30 0.30 3.25		First floor)	
			Girth 4/300 = 1 200	
6/	1.20 2.70		Formwork to sides of isolated columns, regular shape height n.e. 3 m (No. 6)	11.20.1
8/	1.20 3.25		Ditto height n.e. 4.5 m (No. 8)	11.20.1

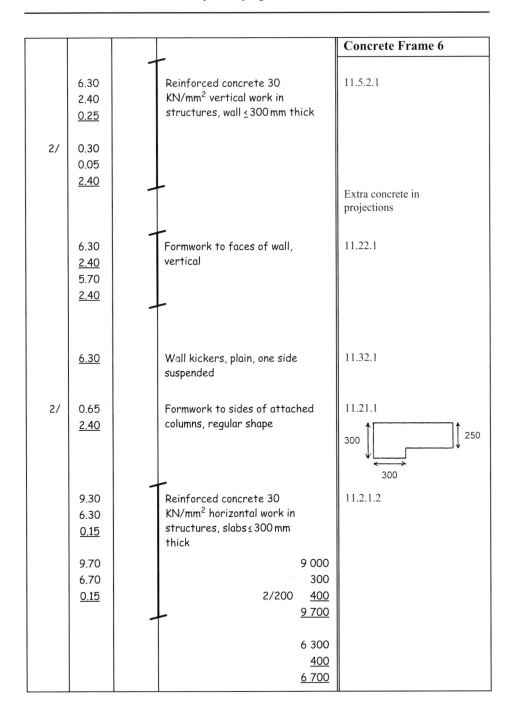

				Concrete Frame 6
	6.30 2.40 0.25		Reinforced concrete 30 KN/mm^2 vertical work in structures, wall ≤ 300 mm thick	11.5.2.1
2/	0.30 0.05 2.40			Extra concrete in projections
	6.30 2.40 5.70 2.40		Formwork to faces of wall, vertical	11.22.1
	6.30		Wall kickers, plain, one side suspended	11.32.1
2/	0.65 2.40		Formwork to sides of attached columns, regular shape	11.21.1 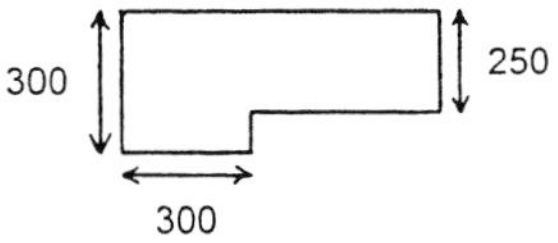
	9.30 6.30 0.15		Reinforced concrete 30 KN/mm^2 horizontal work in structures, slabs ≤ 300 mm thick	11.2.1.2
	9.70 6.70 0.15		9 000 300 2/200 400 9 700 6 300 400 6 700	

				Concrete Frame 7
	9.30 6.30 9.70 6.70		Power float surface of concrete	11.9.1
2/3/ 2/3/	2.70 0.30 0.30 5.70 0.30 0.30 5.70 0.30 0.30 6.30 0.30 0.30		Reinforced concrete 30 KN/mm^2 horizontal work in structures, slabs ≤ 300 mm thick	Columns at wall only measured to underside of beam
3/	2.70 5.70		Formwork to soffit of horizontal work, propping ≤ 3.00 m high & Ditto n.e. 4.5 m high	11.15.1.1
6/ 6/	2.70 0.90 5.70 0.90		Formwork to sides and soffits of square attached beams propping ≤ 3.00 m high	11.18.1.1

				Concrete Frame 8
	2.70 1.05 5.70 1.05 6.30 1.05		Formwork to sides and soffits of attached beams, irregular shape including edge of slab propping ≤ 3.00 m high 900 150 1 050	11.18.2.1
	32.00		Formwork to soffit of horizontal work, projecting slab n.e. 300 mm wide, propping to ground 7.5 m high 9 700 6 700 2/16 400 32 800 Less 4/200 800 Centre line 32 000	11.15.1.3
	32.80		Formwork to edge of horizontal work < 500 mm high	11.14.1
	30.80 0.10 0.15		Reinforced concrete as before, vertical work in structures, upstand 32 000 Less 4/200 800 31 200 Less 4/100 400 Centre line 30 800	11.2.1.2
2/	30.80		Formwork to sides of upstand beam	11.19.1

Chapter 15
Structural Steelwork

The measurement of structural steel can generally be classified as either:

- Framed – framing and fabrication
- Framed – permanent erection
- Isolated members. An isolated member is one which is not part of a frame and would cover work such as isolated beams resting on padstones.

The measurement of the framing is required to be split into two items, with the fabrication being kept separate from the erection, the item for erecting all of the steelwork giving the total weight involved, with the unit of measurement being tonnes.

The weight of the steel supplied to site may be greater than the calculated weight due to the manufacturing process. This is because of the rolling margin for which no allowance is made.

The supply and fabrication of the sections are generally measured by weight, and the structural function needs to be stated; therefore, you would have separate items for items such as columns, beams, bracings, trusses and so on.

In each of the classifications, you need to have separate items for every weight class.

- Weight less than 25 kg per linear metre
- Weight between 25 and 50 kg
- Weight between 50 and 100 kg
- And so on in 50 kg increments.

Willis's Elements of Quantity Surveying, Twelfth Edition. Sandra Lee, William Trench and Andrew Willis.

Published 2014 by John Wiley & Sons, Ltd.

Fittings

The different steel members are joined together by an assortment of plates, brackets, angles and the like. These are called fittings, and they are measured separately by weight in a composite item of fittings for the framed members or the isolated members.

There are two components to a connected joint in steelwork generally:

- A fitting – for example, a cleat which is usually a short length of steel section, either an angle or channel which acts as the connecting agent and is measured as described in this chapter.
- A fixing – this could be rivets, bolts or distance pieces.

The fixing devices are given below.

- Black bolts – forged from round stock with only the threads machined, and used in clearance holes about 1 mm larger than the diameter of the bolt. These are deemed included in the weight of structural members.
- Turned bolts – machined on shank and under head to fit holes with a very small clearance. These are enumerated as 15.11.1.
- Friction grip bolts – high-tensile steel that is tightened to a predetermined tension so that the load is carried by friction; nowadays, this is often used instead of site riveting. Measured as turned bolts.

Holes for bolts and so on are deemed included in the contractors' prices.

Fixing bolts for isolated members are measured where they are required for fixing to other elements and would be enumerated.

Offsite and onsite surface preparation and treatments are measured in square metres.

The measurement of steel framed structures is no more difficult than that of any other type of construction, but, similar to other areas, the following of a logical approach is essential as it is easy to make errors in counting or to omit whole sections of the work.

One approach would be to follow an order of dealing with all work on a particular floor before commencing the next floor. By following this order, there is less chance of something being missed. Within each floor, you might follow the following order.

Main order
- Columns
- Main beams
- Secondary beam

- Filler beams
- Compound trusses
- Purlins
- Rails
- Braces
- Struts

Main member measurement order	*Fittings item*
• Columns (full length taken)	Base plates Splices Caps Cleats
• Beams	Shelf angles Cleats Gusset plates Angles
• Truss main members	
• Struts	
• Ties	

Looking at this order, it will be seen that the columns are given priority of measure, followed by the beams and so on. The fittings order extends the measure of the individual members of the first order to deal with the constituent parts. Splay cuts to members or plates are not measurable, but the members would be measured the maximum length and plates measured to the nearest rectangle of metal.

The measurement of concrete casing to beams and columns would be measured under 11.2.2.2 and 11.5.2.1, and the associated formwork as 11.17.1.2 and 11.20.1.2.

Example 13

Structural Steelwork

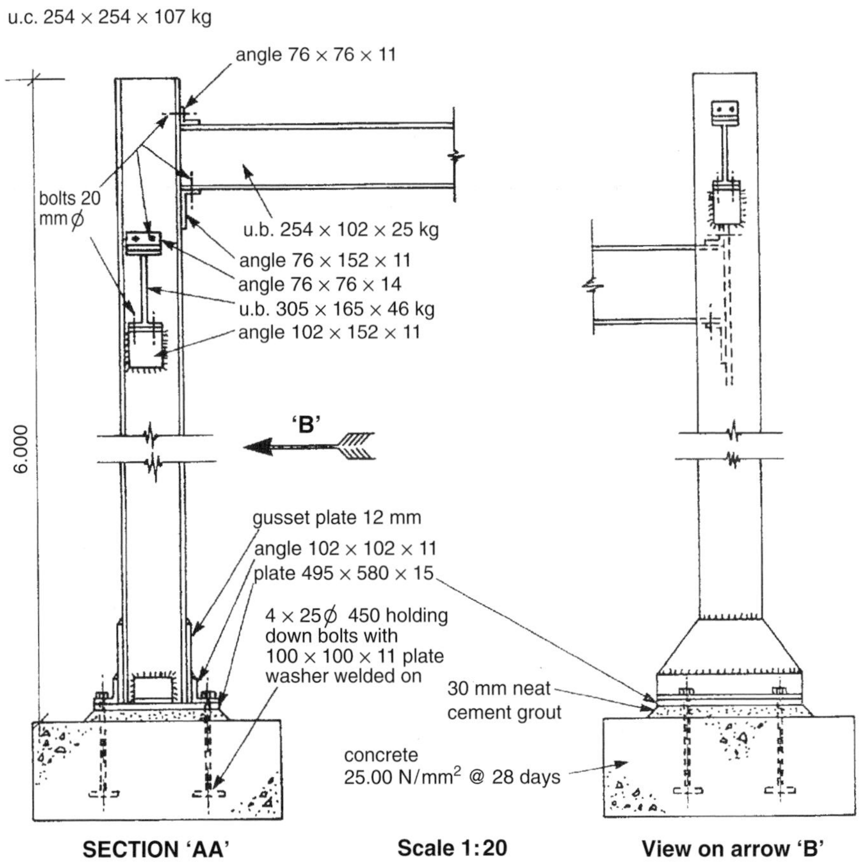

Fig. 37

(continued on following page)

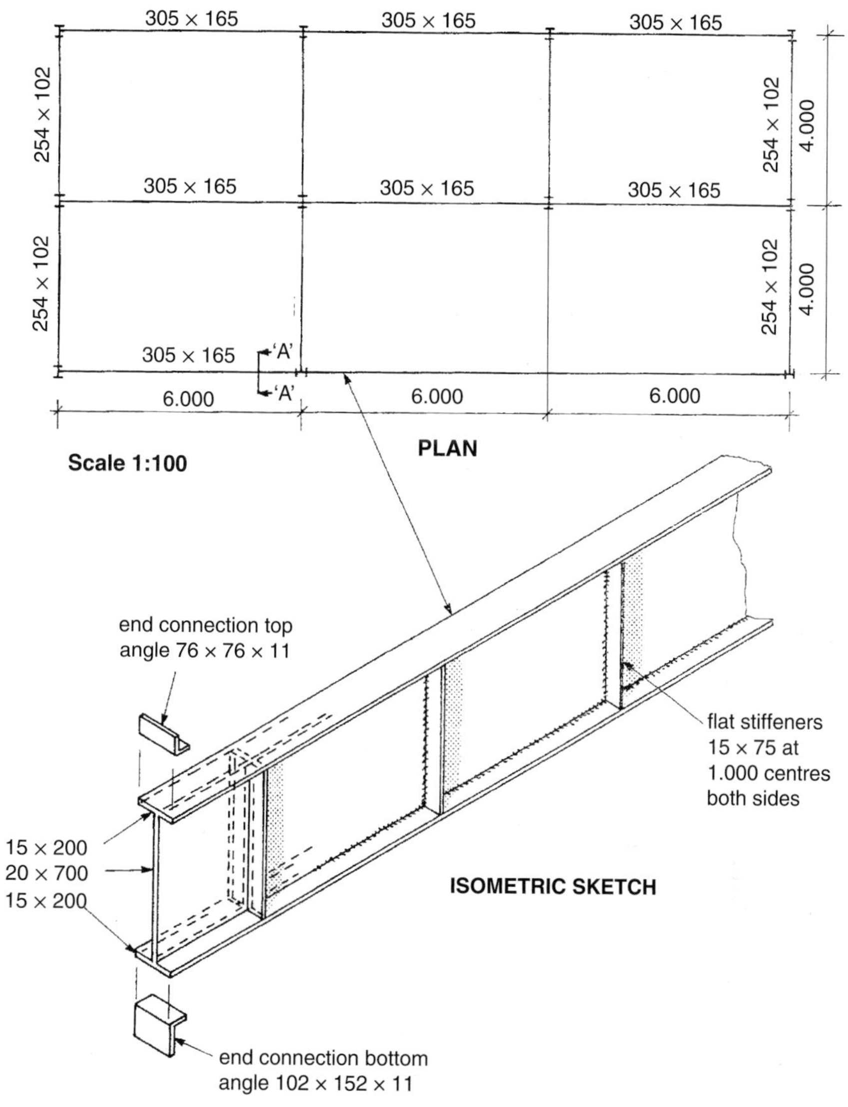

Fig. 37
(continued)

Example 13

Taking-off list	**NRM2 reference**
Framing fabrication:	
Columns	15.1.2.4.1
Fittings: plate	15.5.1.1
angle	
Holding down bolts	15.10.1.1
Grout under base plate	11.42.1.1
Beams	15.1.2.2.2
	15.1.2.1.2
Fittings: angle	15.5.1.1
Built up girder	15.1.3.3.8
Fittings: angle	15.5.1.1
Erection of fabricated steel	15.2

Surface treatment may be required and would normally be measured with the steel frame element

				Steelwork 1
			Preamble notes would be written here	Location drawings to be supplied
			Fabricated framing with factory-welded joints and bolted site connections	
			Column 254 x 254 Precise height 6 000 less grout 0 030 less plate 0 015 5 955	
11/	5.96		Framing, fabricated columns Lengths over 1 m not exceeding 9 m, weight 100–150 kg/m x 107 kg/m = kg	15.1.2.4.1
11/	0.50 0.58		Framing, fabricated allowance for fittings (Weight up the following) 15 mm plate x Kg/m^2 = kg	15.5.1.1 Fittings are grouped together, and these items will be collated with all other fittings to give one bill item. 12 mm plate measured rectangular
11/2/	0.58 0.30		12 mm gusset x Kg/m^2 = kg	
11/2/	0.58		102 x 102 x 11 mm angle x Kg/m = kg	
11/2/	0.15			

				Steelwork 2
7/2/	0.17		102 x 152 x 11 mm angle x Kg/m = kg	For 305 x 165 mm beam
	0.10		76 x 152 x 11 mm angle x Kg/m = kg	For 254 x 102 mm beam
			(End of weighted fittings)	
11/	1		Framing, fabrication holding down bolt assemblies comprising, 4 No. 25 mm diameter m.s. bolts 450 mm long with nut/washer and 100 x 100 x 11 mm plate washer welded on and setting in concrete base with all necessary f/w and wedge and pin up base plate with steel wedges.	15.10.1.1
			&	
			Grout stanchion base with cement grout 25 mm thick	11.42.1.1
			Beam 305 mm length 6 000 Less 2/0.006 0 012 5 988 length 6 000 less $\frac{1}{2}$ column 0 127 less steel 0 006 5 867	Note that the columns at the end of the plate girder are turned through 90°, and therefore one 305 mm beam and two 254 mm beams are different lengths from the others.

				Steelwork 3
6/	5.99		Framing fabrication beams in lengths over 1 m n.e. 9 m long weight 25-50 kg/m	15.1.2.2.2
	5.87		x 46 kg/m = kg	
			beam 254 mm	
			length 4 000	
			less 127/2 0 254	
			3 746	
			length 4 000	
			less 0 133	
			3 867	
5/	3.75		Ditto not exceeding 25 kg/m x 25 kg/m = kg	15.1.2.1.2
2/	3.87			
			Fittings as before (weight up the following)	15.5.1.1
7/2/	0.17		76 x 76 14 mm angle x kg/m = kg	Top angle 305 mm beam
7/2/	0.10		76 x 76 11 mm angle x kg/m = kg	254 mm beam
			(End of weighted fittings)	
			girder length	
			12 000	
			less $\frac{1}{2}$ width col x 2 0.254	
			11.746	

				Steelwork 4
			Framing, fabrication in built-up girder exceeding 9 m long, 25 - 50 kg/m comprising plate web, flanges and stiffeners	15.1.3.8.6
2/	11.75 0.20		Weight up the following:	Flange
2/11/	0.70 0.08		15 mm plate x kg/m^2 = kg	75 mm stiffeners @1 m centres, = 12 less one
	11.75 0.70		20 mm plate x kg/m^2 = kg (end of built up girder)	
2/	0.20		Fittings as before Weight up the following 76 x 76 11 mm angle x kg/m = kg	15.5.1.1
2/	0.20		102 x 152 11 mm angle x kg/m = kg (end of weighted fittings)	
	****		Framing, erection of permanent structure 18 x 8 x 6 m with bolted site connections	15.2 * Total all weighted items = tonnes

Chapter 16
Plumbing

Subdivision

When measuring plumbing, it is particularly important to follow a logical sequence of taking-off in order to be sure that no part is missed. Frequently, particularly on a small domestic installation, the only information shown on the drawings is the location of sanitary appliances. If this is the case, then the measurement of the appliances is fairly straightforward and forms a logical start.

After the sanitary appliances have been measured and possibly coloured in on the drawings, it is then easier to decide on a pipework layout for both wastes and supplies. Sizes of waste pipes are dictated by the size of the waste fitting from the appliance (e.g. wash basins 32 mm and baths and sinks 38 mm). Supply pipework sizes for a small installation should not be too difficult to assess. The rising main is usually in 15 mm pipework, and the down feeds from the cistern in 28 or 22 mm, reducing to 15 mm for the individual feeds except for baths, which require 22 mm. Adequate isolating and drain-off valves should be included in the system to enable sections to be isolated and drained.

Gatevalves and ballvalves do not restrict the flow of water when fully open and should be used on the low pressure distribution part of the system. Some water supply companies still require an indirect system, with a drinking feed to the kitchen sink taken off the rising main and the other cold feeds coming from a storage tank. The capacity of the cold water tank is given as either nominal (i.e. filled to the top edge) or actual (i.e. filled to the working water line). The requirements of water supply companies vary considerably in required capacities, but this should be at least 112 L actual for storage only, rising to 225 L for storage and feeding

Willis's Elements of Quantity Surveying, Twelfth Edition. Sandra Lee, William Trench and Andrew Willis.

Published 2014 by John Wiley & Sons, Ltd.

a hot water system. The cold water tank may have to be raised to provide adequate pressure and flow, particularly for showers, and the roof construction may have to be strengthened to support the additional weight. The inlet to the tank is controlled by a ballvalve, and the outlet should be opposite the inlet to avoid stagnation of water. An overflow pipe with twice the capacity of the inlet should be provided. An alternative installation, where permitted, is for all appliances to be fed directly from the incoming main supply, without the need for a cold water tank.

Before attempting to measure a plumbing installation, trade catalogues depicting fittings available for the specified pipework should be obtained for reference. A selection can then be made of suitable fittings for connections to various appliances. A diagrammatic layout of the plumbing when provided is often not to scale and drawn in two dimensions. When measuring from such a diagram, one has to visualise the layout in three dimensions and to relate pipe runs to the structure. This will enable realistic lengths of pipes to be measured and the correct number of bends to be taken. Sometimes, when measuring copper pipes, it is difficult to decide whether to take *made* bends (i.e. the pipe bent to form the bend) or fittings. Generally, made bends should only be taken for minor changes in direction of pipes or on short lengths. It should be remembered that long lengths of pipes with made bends may be impossible to install. For the measurement of installations without detailed information, the following is suggested as a suitable order:

Sanitary appliances	(a)	Sanitary appliances, including taps, traps, brackets, waste and overflow fittings
Foul drainage above ground	(b)	Wastes, overflow pipes, soil and ventilating pipes, including ducts
Cold water installation	(c)	Connection to supply company's main, supply to boundary of site and meter/stopvalve pit, reinstatement of highway
	(d)	Supply in trench from boundary to building, stopvalve and rising main to the cold water tank
	(e)	Branches from rising main, including exterior taps and non-return valves
	(f)	Cold water tank and lid, including bearers, overflow and insulation
	(g)	Cold down services

Hot water installation	(h)	Feed from cold water tank
	(i)	Boiler, flue, controls and work in connection
	(j)	Cylinder and primary flow and return pipes
	(k)	Secondary circulation, expansion pipe and branch services
Generally	(l)	Casings
	(m)	Chlorination, testing and so on

Note: Insulation of pipes and builder's work in connection (e.g. chases, holes and painting) should be taken after each subdivision.

An alternative approach to measurement is to follow the flow of water from the water main, through the building to the sanitary appliances and discharging into the drains. This is a more logical approach and would probably be adopted if the layout of the whole system is shown on the drawings. The main divisions shown here could still be used but in a different order.

As far as presentation in the bill is concerned, plumbing work has to be classified under headings indicating the nature of the work. For a simple domestic type installation, these would be as follows:

(a) Sanitary appliances
(b) Foul drainage above ground
(c) Cold water
(d) Hot water
(e) Sundry builder's work in connection with services
(f) Testing and commissioning the drainage or water system.

Note: For small-scale installations, (c) and (d) may be combined.

The builder's work in connection section may be billed either under a heading at the end of each appropriate work section or after the installation measurement.

When taking-off, one obviously has to keep in mind these bill divisions, but by following the suggested order of measurement given earlier, the necessary sections will be automatically produced.

Sanitary appliances

If sanitary appliances are specified fully, then they are enumerated and the description should include the type, colour, size, capacity and method of fixing, including details of supports, mountings and bedding and pointing. Frequently, a catalogue or building supply reference is used for part of the description, but care must be taken to define any alternatives available. If full details are not available, then a PC sum may be included for the supply of the appliances, an item included for contractor's profit and fixing measured as enumerated items. Descriptions should make it clear whether items such as taps, traps, overflow assemblies and bath panels are included with the appliance. Small items such as towel rails, mirrors and soap dishes must not be overlooked. Any builder's work such as tile splashbacks, bearers, backboards, painting and similar items necessary for the installation should be measured at this stage.

Foul drainage above ground

Included in this section is the measurement of waste pipes, overflow pipes, anti-syphonic pipes and soil and ventilating pipes. Some appliances, such as water closets (WCs), have traps built in, and some, such as wash basins, have integral overflows. Traps and other pipework ancillaries are enumerated with a dimensioned description and the method of jointing stated, although cutting pipes and jointing materials are deemed to be included. Nowadays, most waste and soil pipes are specified to be in plastic, and their description should state whether they have ring seal or solvent-welded joints and the type and spacing of pipe supports. Pipes are measured linear over fittings, and joints in the running length (i.e. jointing straight lengths of pipe together) are deemed included. The nominal size of pipes has to be given; copper and plastic are usually described by their external diameter, and cast or spun iron and mild steel by their nominal bore. Straight and curved pipes have to be classified separately, and in the case of the latter the radius should be stated. Fixing the pipes to special backgrounds is given as follows:

- To timber, including manufactured building boards
- To masonry, which is deemed to include concrete, brick, block and stone
- To metal
- To metal-faced material
- To vulnerable materials, which are deemed to include glass, marble, mosaic, tiled finishes or similar.

The location of the pipework needs to be included in the description, whether it is in roofs; high or low level in plant rooms, risers or service ducts; or high or low level on floors or in trenches.

There are two alternatives for the measurement of fittings:

(a) Unless measured separately, all fittings are deemed included.
(b) Alternative 1: Fittings, such as bends and tees, are enumerated and the type, material, nominal diameter and method of jointing should be included in the description. Special joints and connections to different pipes and ancillaries are enumerated, stating the method of jointing. Testing the foul drainage is given as an item, giving details of the tests and any attendance required.

General builder's work in connection with an installation is included as an item, along with marking the position of holes, mortises and chases. Pipe and duct sleeves are enumerated with the size and type stated. Painting pipes and radiators, and other painting services, are measured linear to pipes not exceeding 300 mm girth and superficial to those exceeding 300 mm girth. The measurement of overflow pipes to flushing cisterns must not be overlooked; these are measured in the same way as waste pipes.

Cold water

The measurement of the cold water installation will start invariably with the connection to the supply company's main; this would in all probability be included as a provisional sum. It is necessary to check with the company the extent of the work that they will carry out. Often included with the connection is the pipework to the meter/stopvalve pit at the boundary and making good the highway. The meter/stopvalve is required to be located on the pavement or just inside the boundary. Pipes and fittings are measured in the same way as described for waste pipes. Stopvalves, gatevalves and ballvalves are defined as pipework ancillaries, enumerated and described, and their method of jointing stated. Storage tanks are enumerated as general pipeline equipment and are described, including the size and capacity. Overflows to tanks should be taken at this stage.

Insulation to pipelines is measured linear and described, including the thickness of the insulation and the nominal size of the pipe. Working insulation around ancillaries is enumerated as extra over the insulation. Insulation to equipment is either measured superficial (on the surface of the insulation) or enumerated giving the overall size. In the former case, working around ancillaries is enumerated and in the latter case can be included in the item description. Excavating trenches for services are measured linear, stating the average depth in 500 mm stages. Earthwork support, consolidation, backfilling and disposal are deemed to be included in the trench item. Meter/stopvalve chambers and boxes are each enumerated and described. Underground ducts are measured as linear items, stating the type, nominal size and method of jointing and whether they are straight or curved. The remainder of the builder's work is measured as described in this volume.

Hot water

Traditional domestic hot water systems, apart from the pipework, have three main components – the boiler, the cylinder for storage and the cold feed storage. Suitable pipe sizes would be 28 mm for the primary flow and return between the cylinder and boiler and for the cold feed to the cylinder. The hot water distribution from the cylinder would be 28 mm, reducing to 22 mm for the vent, 15 mm for sink or basin supplies and 22 mm for the bath. These sizes should be regarded as minima; sizes would be increased for a larger number of draw-off points. When an indirect heating circuit is included in the system, then either a self-venting cylinder or a separate expansion and feed tank have to be provided. Whilst heating installations are considered to be beyond the scope of this book, it is worth mentioning that a separate bill heading of low-temperature hot water heating (small-scale) would have to be introduced. Boilers and cylinders are enumerated and described under the rules for equipment; the description should include, as appropriate, the type, size, pattern, rated duty, capacity, loading and method of jointing. The remainder of the work is measured as described here.

Example 14

Internal Plumbing

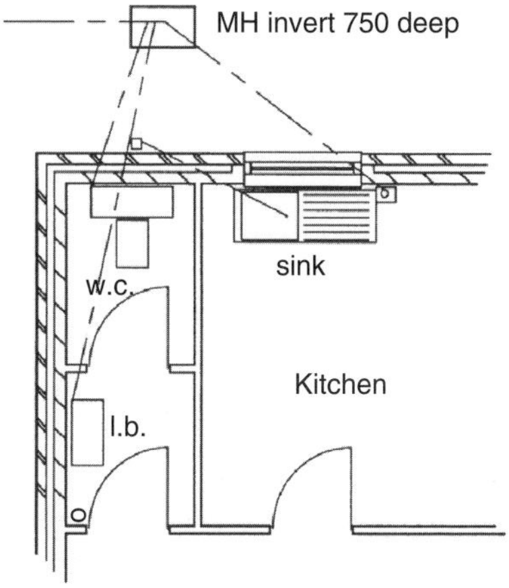

GROUND FLOOR PLAN

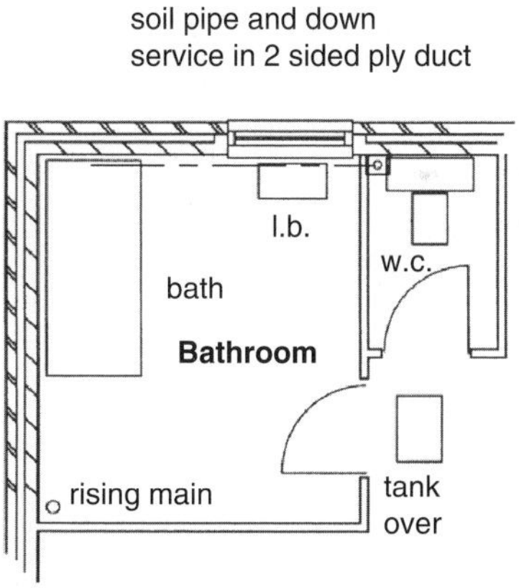

FIRST FLOOR PLAN

Scale 1:100

Fig. 38

Example 14

Taking-off list	**NRM2 reference**
Sanitary appliances: prime cost sum	
fixing	32.2.2
Sink and drainer	32.2.2
Accessories support battens	16.3.2.1
Foul drainage: 110 mm pipes	33.1.1
joint to drain	Not measurable
bends	33.2.1
32 mm traps	33.2.1
40 mm ditto	
32 mm pipe	33.1.1
40 mm pipe	
fittings	33.3.1
joints to stack	Not measurable
Overflows: pipes	33.3.1
fittings	33.3.1
Marking holes etc.	41.2.1
Testing	33.10.1
Builder's work in connection	
Cold water supply:	
pipes	38.3.1
fittings	Deemed included
stopcock	38.5.1
storage tank and ballvalve	38.1.1
overflow	38.3.1
fittings	Included
Marking holes etc.	41.2.1
Testing	38.17
Commissioning	38.18
Builder's work in connection	
Insulation to cold supply: pipe	38.9.2
tank	38.9.1
Painting pipes	29.6.1
Tank bearers	16.3.1
Backboards	16.4.1

				Internal Plumbing 1
			Sanitary appliances	
			Include the prime cost sum of £__________ for sanitary appliances	32.2.2 Assuming that the specification has not yet been decided.
			Profit %	
			Fixing and assembling the following, including bedding **taps** and waste fittings in mastic and bedding and pointing at abutments with walls in flexible sealant	
2/	1		White glazed vitreous china wash basins size 460 x 405 mm with a pair of chromium-plated pillar taps, waste fitting, plug and chain and pedestal, including plugging and screwing brackets to masonry and bedding pedestal to floor and basin in mastic.	32.2.2 As much information as possible should be given to allow the estimator to accurately assess the labour involved.
			First Floor	
	1		White glazed vitreous china WC suite with a low-level 9 L plastic cistern and ballvalve. Plastic flush pipe, seat and cover, including plugging and screwing cistern brackets to masonry and screwing pan to timber and bedding in mastic.	32.2.2

				Internal Plumbing 2
			Ground Floor	
	1		Ditto but screwing pan to masonry and bedding in mastic.	The ground-floor WC is fixed to the concrete floor.
	1		White reinforced acrylic bath size 1 700 x 700 mm with a pair of chromium plated pillar taps, overflow fitting and flexible tube, waste fitting, plug and chain, moulded front and end panel including screwing adjustable bath feet to timber, plugging and screwing wall brackets to masonry and screwing panels to timber	
	1		Stainless-steel double-bowl, double-drainer sink unit size 2000 x 600 mm with chromium plated mixer **tap,** overflow fitting and flexible tube, waste fitting, plug and chain, including setting in timber base unit (measured) and fixing with screws	The timber base unit has been taken with the kitchen fittings.
	1		Chromium-plated toilet roll holder including plugging and screwing to masonry	Any other items such as mirrors would be included here.
			End of fixing and assembling	

				Internal Plumbing 3
			Bath panel 1 700 700 2 400	
	2.40 0.60		Sawn softwood framed battens overall width 600 mm, 38 x 38 mm spaced **at** 600 mm c/c	16.3.2.1 Bath panel framing and other ancillary work for sanitary appliances are taken here.
			Foul drainage above around soil and vent pipe Ground floor 2 600 First floor 2 600 Floor 250 5 450 Roof space 600 Above roof 450 1 050	
	5.45		Pipes, straight uPVC, 110 mm with ring seal joints and socket pipe clip plugged and screwed to masonry in duct	33.3.1.1 Pipes in ducts must be identified separately.
	1.05 0.80 0.30		Ditto but not in duct	Soil pipe in roof. Branch first-floor WC.
	1		Ancillaries, 110 mm dia. terminal solvent welded	33.2.1 In top of soil stack

				Internal Plumbing 4
2/	1		Extra over 110 mm uPVC pipe for bent WC connector	33.3.1 Joints to drain soil stack and ground-floor WC deemed included.
	1		Ditto for 110 mm boss branch, including ring seal joint	33.3.1 First floor level for WC
	1		Ancillaries, 32 mm polypropylene tubular swivel Y trap with 76 mm seal	33.2.1 Washbasin traps
	1		40 mm ditto	Sink
2/	1		40 mm ditto bath trap with 32 mm overflow bend	Bath

				Internal Plumbing 5
	0.60 1.50		Pipes, straight (ground floor) MuPVC 32mm (first floor) With solvent welded joints and clips plugged and screwed to masonry	33.1.1
	3.50 0.80		Ditto 40 mm	33.1.1
2/	2		Extra over 32 mm MuPVC pipe for fittings two ends	33.3.1 Fittings to pipes n.e. 65 mm are described according to number of ends.
	1		40 mm ditto	Bath and sink
2/	1		Extra over 110 mm uPVC pipe for 32 mm boss branch with ring seal joint	Boss for pipe to first floor wash basin
	1		Ditto 40 mm boss branch	Bath waste

				Internal Plumbing 6
			Overflows to cisterns	
2/	1.00		Pipes, straight uPVC 22 mm overflow with solvent welded joints and clips plugged and screwed to masonry	33.1.1
2/	2		Extra over ditto for fittings two ends	33.3.1
	Item		Marking the positions of holes, mortises and chases in the structure for the foul drainage above ground	41.2.1
	Item		Testing and commissioning the foul drainage above-ground installation	33.10.1

				Internal Plumbing 7
			Builder's work Soil and vent pipe	
	1		Aluminium patent weathering slate 457 x 400 mm with moulded rubber sealing cone for 110 mm uPVC pipe handed to others for fixing & Fixing only ditto (by roofer)	18.4.1.8
			End of builder's work	

				Internal Plumbing 8
			<u>Piped supply systems: water supply (cold water)</u>	It is assumed that the incoming main has been measured elsewhere.
			Copper tubes to be to BS 2871 Part 1, half hard, table x	
			Fittings for copper tubes to be to BS 864 Part 2 compression Type A (non-manipulative) and are deemed included.	Common headings inserted here to avoid repetition
			Copper tube to be fixed with standard saddle band type clips at 1 200 mm centres fixed with brass screws	
			Ground floor 2 600 **First** floor 2 600 Floor <u>250</u> <u>5 450</u>	38.3.1
	<u>5.45</u> <u>6.40</u>		Pipes, straight copper 15 mm clips plugged and screwed to masonry	Rising main and branch to sink

				Internal Plumbing 9
			To sink 3 200 2 500 700 6 400 4 000 Clg 100 Vent to cistern 500 4 600	
	4.60		Ditto but clips screwed to timber	In roof It is at the surveyor's discretion whether to measure the fittings out or to deem them to be included.
			Stop valve at entry	
	1		Combined high-pressure screw-down stop valve and drain tap as BS 1010 and 2879A. Compression joints to copper tube	

				Internal Plumbing 10
			Tank in roof	
	1		Water storage tank to BS 4213 polythene 225 L capacity, including lid, perforations for one 15 mm and three 22 mm pipes and backing plates including ball valve, high pressure; BS 1968 Class A PVC float 15 mm inlet fixed to polythene tank and connection to copper pipe, including straight coupling	38.1.1 Water storage tank in roof
			Overflow	
	3.50		Pipes, straight uPVC 22 mm overflow with solvent welded joints and clips screwed to timber	38.3.1
	Item		Marking the position of holes, mortises and chases in the structure for the cold water installation	Fittings deemed included, and connections deemed included. 41.2.1

				Internal Plumbing 11
	Item		Testing the cold water service	38.17
	Item		Commissioning the cold water installation	38.18
			Builder's work in connection with plumbing	
				Holes in new structure are not measureable.
	5.45 6.40		Prime and two coats gloss paint copper services, isolated surfaces not exceeding 300 mm girth	29.6.1.1 Painting exposed pipes

				Internal Plumbing 12
			Tank bearers	
3/	1.50		Individual battens 50 x 100 mm sawn softwood pressure impregnated	16.3.1
			Tank platform	
	1		Backboards etc. 1 500 x 1 250 chipboard 19 mm BS 5669 flooring grade	16.4.1.1
			Insulation to cold water supply	
	4.60		Insulation 20 mm thick glass fibre sectional to 15 mm diameter copper pipes, including metal bands	38.9.2 Working around finings is deemed included, but working around ancillaries is measured.
	1		Insulation 25 mm thick expanded polystyrene boards to sides and top of tank 225 L capacity, including securing with bands and cutting around pipes.	38.9.1
				Note: The down services have been omitted from this example as it is mainly repeating items already measured.

Chapter 17
Drainage

Subdivision

The measurement of drainage may be divided into the following sections:

- Manholes or inspection chambers
- Main drain runs and fittings between manholes
- Branch drain runs and fittings between outlets and manholes
- Accessories (e.g. gullies)
- Sewer connection
- Land drains
- Testing.

If the foul water and the surface water drainage are separate systems, then usually each would be measured independently using the above sequence. When measuring surface water drainage, there must be liaison with the person measuring the roads and pavings in order to ascertain the position of gullies and channels. Similarly, there will have to be consultation with the person measuring the roofs in order to ascertain the position of rainwater pipes and to decide who will measure the connections if required. When measuring foul water drainage, there will have to be similar discussion with the person measuring the plumbing in order to find out the position of soil pipes and outlets and again to establish who is to measure the connections. The opportunity should be taken during these discussions to check that the drawings show drain runs leading from all service outlets.

Willis's Elements of Quantity Surveying, Twelfth Edition. Sandra Lee, William Trench and Andrew Willis.

Published 2014 by John Wiley & Sons, Ltd.

Manholes

In most cases, the position of manholes will be shown on the drawings together with invert levels. If the existing and finished ground levels adjacent to the manhole are not shown, then these will have to be ascertained from the site plans, possibly by interpolation.

Prior to the measurement of manholes, it is wise to prepare a schedule of similar format to that shown in Fig. 39. If information on the sizing of manholes is not available, then the following guide may be useful:

- For inspection chambers up to 900 mm deep, the minimum internal size should be 700 mm wide × 750 mm long, allowing up to two branches on each side.
- For manholes over 900 mm and up to 3300 mm deep, the minimum size should be 750 mm wide × 1200 mm long, allowing 300 mm in the length for each 100 mm branch and 375 mm for each 150 mm branch.
- For manholes up to 2700 mm deep, a cover of size 600 × 600 mm should be provided, increasing to 600 × 900 mm for deeper manholes.

MANHOLE SCHEDULE

Nr	DIAGRAM	INTL SIZE	DEPTH EXCAVATION	SIZE 150 CONC BASE	SIZE 150 RC COVER	COVER & FRAME
1	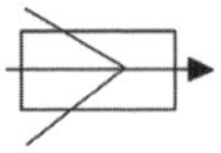	563 × 675 mm	675 Chan 50 Base 150 875	563 675 750 750 1313 × 1425	563 675 430 430 993 × 1105	457 × 457 CI BS EN 124
2 etc.						

Nr	WALLS ONE BRICK HEIGHT	GIRTH	CHANNEL	BRANCH BENDS	BUILD IN ENDS	REMARKS
1	675 + 50 = 725 − 150 575	563 675 2/1238 = 2476 4/215 = 860 3336	100 mm Str 675 mm long	Two 100 mm	Four 100 mm	
2 etc.						

Fig. 39

- For deep manholes, an access shaft can be constructed at the top to within 2 m of the benching.
- When the depth of the manhole exceeds 900 mm, step irons should be taken at 300 mm intervals.

Manholes, inspection chambers and the like are enumerated with a detailed description of the construction and internal dimensions, giving the depth from the cover level to the invert level in 250 mm stages.

Drain runs

Once the manholes have been measured, it is a comparatively simple matter to measure the main drain runs between. Firstly, a schedule should be prepared on the lines of that shown in Fig. 40. This schedule has been prepared for the drain runs in Example 15, and it could have replaced many of the waste calculations. The depth of the drain excavation will usually be the same as that for the manhole at each end, although a small adjustment may have to be made if the bed below the manhole is of different thickness to that below the drainpipe. The depths at each end of the drain run will be averaged, and the depth classified in 500 mm stages. Drain runs including the pipes are measured as linear items stating the type and nominal size of the pipe. If multiple pipes are included in the trench, then this should be stated. Earthwork support, treating bottoms, filling and disposal are all deemed to be included with the trench item.

Bedding and surrounds for pipes should be given in the description of the drain run pipes.

The pipe length through the manhole wall, building in the end of the pipe and any bedding or pointing are deemed included. Pipe fittings such as bends are enumerated, and the method of jointing given.

Branch drain runs are measured in the same way as main drain runs, and again a schedule should be prepared before measurement. The important difference is that branch runs are connected to the manhole at one end only. At the other end will be either a gully or a connection to a soil pipe or similar outlet. The depth of the trench at this end will depend upon the size of the gully or rest bend, but for normal circumstances it could be taken as 600 mm. If bends are used, then it must be remembered that the run must be measured to include their length. Gullies and other accessories are enumerated and described, jointing to pipes and concrete bedding being deemed included. Back and side inlets, raising pieces and gratings should be included in the description of the gully. Usually, two bends are required in a pipe coming from a gully as there may have to be a change in direction as well as an adjustment to achieve the correct fall. Circular-top or two-piece gullies may, however, reduce the need for direction change.

The connection to the public sewer, if carried out by the contractor, is enumerated and described. If the connection is carried out by the statutory authority, as is usually the case, then a provisional sum is included to cover the cost of the work. Care must be taken to ascertain the extent of the work done by the authority and to include for the remainder in the measured work. The work frequently involves excavation across the highway, which would require special provisions such as traffic control, lighting and reinstatement. It should also be noted that work beyond the boundary of the site has to be kept separate.

Testing the drainage system is included as an item, stating the method to be used. The extent of testing required should be ascertained from the appropriate authority.

Example 15
Drainage

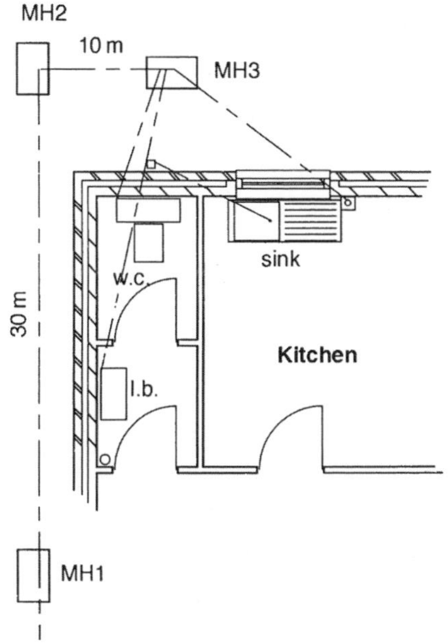

GROUND FLOOR PLAN

Fig. 40

Example 15

Taking-off list	NRM2 reference
Manholes	34.6.1
Main pipe runs	
Excavation for trench	34.1.1
Pipes	34.1.2
Concrete bed and haunch	Not measureable
Pipe fittings	34.3.1
Pipe ancillaries	34.4.1
Branch in building	
Excavation	34.1.1
Cast iron pipe	34.1.2
Concrete bed and surround	Not measureable
Pipe fittings	34.3.1
Trapped gulley	34.4.1
Testing	34.17.1

Trench excavation depth allows for 100 mm bed and 20 mm thickness of the pipe. The length is measured between outside face of manholes.
Formation level has been assumed at 100.00, around house, and then falling away along the main runs.
Pipes are assumed to fall at least 1:40 (therefore over 30 m the drop is 750 mm and over 10 m the drop is 400 mm).
Pipes measured externally to house and up to manhole 1.

Run location	*Pipe size 100 mm*	*Form. level at head*	*Invert level at head*	*Exc depth at head*	*Form. level at foot*	*Invert level at foot*	*Exc depth at foot*	*Ave. dp of exc*	*Exc tr for pipe n.e. 200 mm dia*		
									0.75	1.00	1.25
SVP-MH3	2.10	100.00	99.55	0.57	100.00	99.25	0.87	0.72	2.10		
Gulley-MH3	1.00	100.00	99.55	0.57	100.00	99.25	0.87	0.72	1.00		
WC to MH3	1.60	100.00	99.55	0.57	100.00	99.25	0.87	0.72	1.60		
MH3-MH2	10.00	100.00	99.25	0.87	99.75	98.85	1.02	0.945		10.00	
MH2-MH1	30.00	99.75	98.85	1.02	99.00	98.10	1.02	1.02			30.00
BED	**44.70**										
add for pipe into house	1.20										
into manholes 7 × 215	1.51										
PIPES	**47.41**						**length of trenches**		**4.70**	**10.00**	**30.00**

Head is the shallow end of a run, foot is the deeper end of a run.
To calculate excavation depth at head or foot is (formation level less invert) plus thickness of pipe and bed (130 mm)

Fig. 41

				Drainage 1
			Manholes Internal size 900 x 570 Wall 2/215 430 430 1 330 1 000	Manhole construction is one brick thick walls and 150 mm thick concrete base slab with no spread
			Depth M/H 1/2 3 Invert 900 750 Bedding 25 25 Base 150 150 1 075 925	
	1		Manhole; maximum internal size 900 x 570 mm; depth to invert over 750 but n.e. 1 000 mm deep; constructed with 150 mm thick concrete base (20 N); one brick thick walls in semi-engineering bricks, English bond in cement mortar (1:3), flush pointed one side; 100 mm dia. curved main channel 900 mm long with Nr 2 three-quarter section branch bends; precast concrete cover slab 1 330 x 1 000 x 100 mm thick reinforced with steel fabric to BS 4483 ref A142 with rebated opening 600 x 450 mm finished smooth and setting in cement mortar (1:3)	Manhole 3 34.6.1
	2		Ditto depth to invert over 1 000 n.e. 1 250 mm	34.6.1 Manholes 1 and 2
	3		Galvanised mild steel single seal cover and frame 600 x 450 mm and setting frame in cement mortar and cover in grease.	34.14.1

				Drainage 2
			Pipe runs to M/H 3 Branch depth 450 M/H 3 750 2) 1 200 600 add bed and pipe 130 average depth 730 Main run M/H 3 - M/H 2 Invert depth 750 900 2)1 650 825 add bed and pipe 120 average depth 945 Main run M/H 2 - M/H 1 Invert depth 900 900 2)1 800 900 add bed and pipe 120 average depth 1 020	
	1.60 1.00 2.10 10.00		Drain runs; 100 mm dia. vitrified clay pipes with flexible joints; average depth 500 - 1 000 mm; in situ concrete (20 N) bed and haunching 425 x 100 mm section	34.1.1 If special material is required for backfilling, this would be given in the description.
	30.00		Ditto 1.00 - 1.5 m deep	

				Drainage 3
3/	2		Pipe fittings, 100 mm dia. vitrified clay bend	34.3.1
	1		Accessories, clay trapped gulley with 100 mm outlet, 225 mm raising piece with No. 2 back and side inlets, 100 x 50 mm reducing sockets and galvanized grating	34.4.1
			Branches to gulley	
	1.40 2.20		Drain runs; 50 mm dia. cast iron pipes with spigot and socket joints; average depth n.e. 500 mm; in-situ concrete (20 N) bed and surround 375 x 275 mm section	
2/	2		Pipe fittings, 50 mm dia. cast iron bend	
	item		Testing and commissioning	

Chapter 18

External Works

Particulars of the site

Before the measurement of external works is commenced, a visit to the site should be made to ascertain items to be included or to check the information shown on the drawings. Items which could be checked or taken include:

- Grid of levels and dimensions of the site
- Pavings etc. to be broken up
- Demolition of walls, fences, buildings etc.
- Felling trees and grubbing up hedges
- Preservation of trees and grassed areas etc.
- Any existing services, overhead power lines etc.
- Any turf that may be worth preserving.

In addition to noting these items, consideration could be given at the same time to other matters, such as access to the site, which may have to be drawn to the attention of the tenderers in the preliminaries.

Coverage

The measurement of external works can include several different aspects of work. To give an idea of the coverage a selection of items is listed below:

(1) Site clearance including:
  - Removal of trees, hedges and undergrowth
  - Lifting turf for reuse

Willis's Elements of Quantity Surveying, Twelfth Edition. Sandra Lee, William Trench and Andrew Willis.

Published 2014 by John Wiley & Sons, Ltd.

- Breaking up pavings etc.
- Demolition work.

(2) Temporary works (other than those for the contractor's own convenience) including:
- Roads
- Fencing
- Maintaining existing roads.

(3) Roads, car parks, paths and paved areas including:
- Preparatory work
- Pavings
- Kerbs, edgings, channels
- Drop kerbs
- Steps and ramps
- Road markings
- Surface drainage.

(4) Fencing and walls including:
- Boundary, screen, retaining walls
- Fencing, gates
- Guard rails
- In situ planters.

(5) Outbuildings including:
- Garages
- Substations
- Gatekeepers' offices
- Bus shelters
- Canopies.

(6) Sundry furniture including:
- Bollards
- Seats and tables
- Litter, grit and refuse bins
- Cycle stands
- Prefabricated planters
- Flag poles
- Clothes driers
- Sculptures
- Signs and notices
- Lighting standards.

(7) Water features including:
- Lakes and ponds
- Ornamental and swimming pools
- Fountains.

(8) Horticultural work including:
- Cultivating, topsoil filling, subsoil drainage
- Grassed areas (seeding and turfing)
- Planting trees (tree grids and guards)
- Planting shrubs, hedges and herbaceous plants.

(9) Sports facilities including:
- Playing fields and running tracks
- Tennis courts
- Bowling greens
- Children's play equipment.

(10) Maintenance including:
- Planted and grassed areas
- Playing fields.

(11) External services including:
- Water, gas and electric mains
- Telephone and TV services
- Fire and heating mains
- Security systems.

Whilst the above areas of work have been classified under various headings, the presentation of the work in the bill is a matter for personal preference. NRM2 provides specific measurement rules for roads and pavings, edgings, soft landscaping, fencing and site furniture, and it would be as well to follow these at least.

Site preparation

Trees and tree stumps to be removed from the site are taken as enumerated items, stating the girth measured 1.00 m above ground as 500 mm to 1.50 m, 1.5 to 3 m, and exceeding 3 m. Grubbing up roots, disposal and filling voids are deemed to be included although a description of the filling has to be given. Site clearance is measured metres square and includes for all trees up to 500 mm girth. Lifting turf for preservation is given as a superficial measurement as is the removal of topsoil (depth stated) and the breaking up of existing pavings (type of paving and depth stated).

Excavation

General excavation in connection with external works is measured under the same rules as for general building work. The main problems in the measurement of external works are likely to arise from bulk excavating to reduce levels in uneven ground and from measuring irregular areas. Guidance on calculations for these items is given in Appendix 1. Provided that a proper grid of levels has been taken over the site, it should be fairly simple to find the average depth of excavation.

There may be a difficulty in giving the required depth stages for excavation when depths vary over the site. If this appears to be a problem and would mean drawing many contours representing depth changes, it is suggested that, irrespective of the required depth stages, an average depth is found for the area and this is used for the classification. The contractor should, of course, be informed that this method has been adopted. Frequently it is prudent to divide large irregular areas into sections for measurement; not only will this assist with the calculation of the area but it will also give greater accuracy in the depth measurements and classifications. Furthermore, it may be wise to make a division at the edge of an area of shallow excavation where it becomes deeper. Handling excavated material is generally the responsibility of the contractor but if there are any specified conditions regarding handling, such as the provision of temporary spoil heaps, then these have to be stated.

Roads and paving

Specifications for roadworks frequently refer to guides and recommendations published by the local authority or government department. Concrete roads are measured as cubic items and classified as for slabs. Mechanical treatment to the surface of the concrete is given as a superficial item. Macadam roads are measured superficial, the area taken being that in contact with the base or lineally if less than 300 mm wide. No deductions are made for voids within the area when not exceeding 1.00 m^2. Work is described as being level and to falls, to falls and crossfalls and slopes not exceeding 15° slope, or to slopes exceeding 15°. Forming or working into shallow channels is included but linings to large channels are taken as linear items, stating the girth on face; all labours are deemed included. Gravel paving and roads are measured and classified in the same way as macadam.

Brick and block paving is also measured in a similar manner except that measurements are taken on the exposed face. The SMM includes rules for the measurement of special pavings for sport; these are measured the area in contact with the base and are again classified under the same categories mentioned above. Kerbs, edgings and channels are measured linear, with specials (such as corner blocks or outlets) being enumerated as extra over items.

Walling

Walling in connection with external works is again measured in accordance with the general rules although particular attention needs to be given to the measurement of any curved and battered work.

Fencing

Fencing is measured as a linear item over the posts. There are several specifications for fences and these can frequently be used to assist with descriptions. Posts or supports occurring at regular intervals are included in the description of the fencing but occasional supports such as straining posts are enumerated and described as extra over the fencing. The excavation of post holes, backfilling, earthwork support and disposal of surplus are deemed included but the size and nature of the backfilling have to be stated. Gates are enumerated and ironmongery is also enumerated separately.

Sundry furniture

Most of the items listed above under 'Coverage', section (6), Sundry furniture, are enumerated and supported by a component drawing, dimensioned diagram or reference to a trade catalogue or standard specification.

External services

As far as services are concerned, pipes in trenches have to be kept separately. Special rules apply to the measurement of trenches for services; these are measured as linear items stating the nominal size and type of service. Average depths are stated in 500 mm stages; earthwork support, backfilling and disposal are deemed included.

Example 16

Roads and Paths

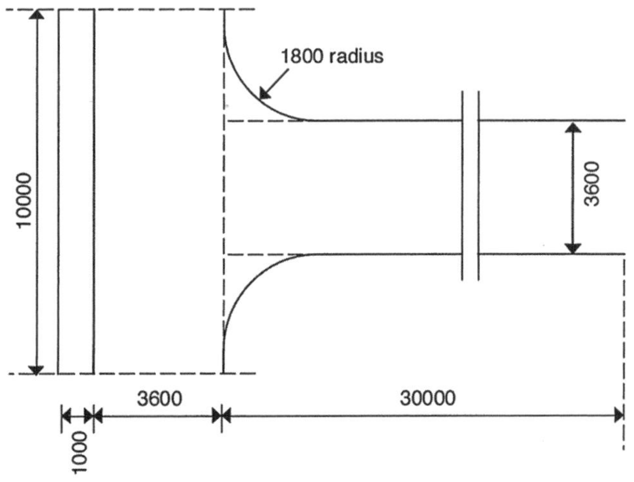

PLAN **Scale 1:200**

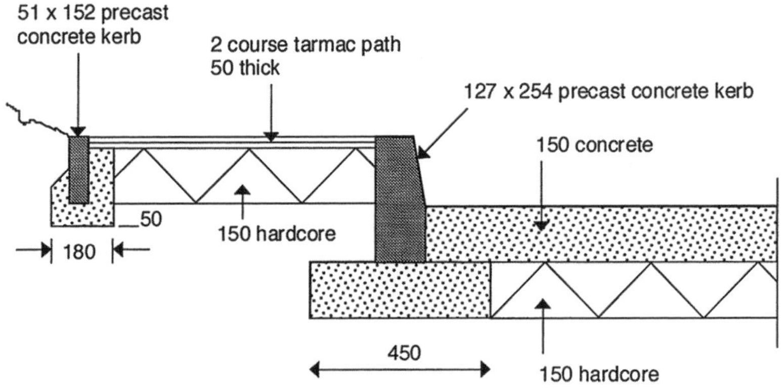

EDGE DETAIL **Scale 1:20**

Fig. 42

Example 16

Taking-off list	**NRM2 reference**
Excavate to reduce levels	5.6.1
Soil disposal	5.9.2
Compact ground	Not measurable
Earthwork support	Not measurable
Excavate trench	5.6.2.1
Filling under pavings	5.12.2.1
Surface treatment	Not measurable
Damp-proof membrane	5.16.2
Concrete bed	11.3.1.2
Surface treatment tamping	Not measurable
Reinforcement	11.37.1
Design joints	11.38.1
Concrete road kerb	35.1.1
Curved kerb	35.1.1.1
Path kerb	35.2.1
Adjustment for bell mouth area	
Tar macadam paving	35.12.1.2
Adjust soil disposal	5.9.2

				Roads and Paths 1
			Road length	
			30 000	
			10 000	
			40 000	
			Width Depth	
			3 600 254	
			Kerb 2/127 254 150	
			Spread 2/162 324 404	
			4 178	
			Path	
			Depth 50	
			150	
			200	
			Path 1 000	
			Less kerb 289	
			711	
			51	
			50	
			101	
			812	
	10.00 0.81 0.20		Excavation. Bulk excavation n.e. 2 m deep	5.6.1
			&	
	40.00 4.18 0.40		Disposal excavated material off-site	5.9.2 The area bounded by the curve is added later, although it could be included with each item here.

				Roads and Paths 2
	10.00 0.18 0.05		Excavation. Foundation excavation n.e. 2 m deep	5.6.2.1
			&	
	40.00 4.18 0.40		Disposal excavated material off-site	5.9.2
			1 000 less 127 79 - 206 794	
	40.00 3.28 10.00 0.79		Imported filling, hardcore blinding bed n.e. 50 mm thick, average 25 mm thick, level and to falls x 0.025 = m^3	5.12.1.1
			&	
			Damp-proof membrane, over 500 mm wide, horizontal, waterproof building paper to BS 1521 grade B1 lapped 225 mm at joints and laid on blinded hardcore to receive concrete	5.16.2.1

				Roads and Paths 3
	40.00 3.60 0.15		In-situ concrete bed (20 N) n.e. 150 mm thick in bays n.e. 40 m^2	11.3.1.2 Tamping surfaces deemed included.
	40.00 3.60		Reinforcement, welded fabric to BS 4483 type A252 weighing 3.95 kg lapped 150 mm at joints	11.37.1
3/	3.60		Designed joint in in-situ concrete n.e. 150 mm deep with pre-moulded impregnated fibre board, the top 15 mm filled with approved compound to BS 2499 and cutting fabric	11.38.1
2/	28.20 10.00 2.80		Precast concrete (granite aggregate) kerb to BS 7263 127 x 254 bedded and jointed in cement mortar and haunched in concrete (10 N), including 450 x 150 mm concrete foundation	35.1.1
2/	2.83		Ditto curved on plan, 1 800 mm radius	35.1.1.1

				Roads and Paths 4
	10.00		Precast concrete edging Fig 10, 51 x 152 mm bedded and jointed in cement mortar and haunched in concrete (10 N), including 180 x 50 mm concrete foundation	35.2.1.1
			Adjustment for bell mouth Area = 3/14 x r x r r = 1 800 mm = 0.695 m^2	
2/	0.70 1.00		Excavating, bulk excavation n.e. 2 m deep x 0.40 = m^3	It is best to show two dimensions to indicate that this is a square metre item.
			&	
			Disposal excavated material off-site	
			&	
			Imported filling, hardcore a.b. x 0.15 = m^3	
			&	
			Damp-proof membrane a.b.	
			&	
			In-situ concrete bed a.b. x 0.15 = m^3	
			&	
			Fabric reinforcement a.b.	

				Roads and Paths 5
	10.00		Two coat Macadam paving to BS 4987, to falls and crossfalls, base course 30 mm thick, wearing course 20 m thick rolled and blinded with grit	35.12.1.2
	0.82			
			1 000	
			less 127	
			50 - 177	
			823	
	10.00		Deduct	
	0.16			
	0.15		Disposal excavated material off-site a.b.	
	59.20			
	0.16			
	0.25		&	
2/	2.83			
	0.16		Add	
	0.25		Filling obtained from excavated material depth n.e. 500 mm, compacted in 150 mm layers	5.11.1.1 Adjustment of filling behind kerb and edging

Chapter 19
Demolitions, Alterations and Renovations

Generally

The work covered by this chapter relates entirely to existing buildings.

NRM2 has separated the rules for the measurement of demolitions from those for alterations. The 'Demolitions' section also includes the following:

- Demolition
- Temporary support of structures
- Temporary works
- Decontamination
- Recycling

The 'Alterations' section includes the following items:

- Works of alteration – spot items
- Removing
- Filling in openings or recesses
- Cutting or forming openings or recesses
- Removing existing and replacing
- Repairing existing structures for the attachment of new work
- Repairing
- Repointing.

The main points to remember when measuring this type of work are firstly to identify carefully the exact location of the work, secondly to describe clearly what has to be done and thirdly to state which new items are included as opposed to being measured separately. To illustrate the

Willis's Elements of Quantity Surveying, Twelfth Edition. Sandra Lee, William Trench and Andrew Willis.

Published 2014 by John Wiley & Sons, Ltd.

last point: when describing cutting a new opening, state whether or not the new lintel and flooring in the opening are to be included in the price. Often, it is advisable to measure and describe the items when on site so that one is aware of the existing conditions. Usually, any locations given should relate to the existing building rather than to the proposed layout, thus enabling estimators to locate them easily when pricing on site.

Surplus materials arising from their removal become the property of the contractor, who is responsible for their disposal. If the employer wishes to retain any of the surplus materials or reuse them in the works, then special mention has to be made of this in the bill. If it is felt that any of the surplus materials may be of use or value, then the contractor could be requested to price their removal with a separate amount for credit, with a proviso that the employer reserves the right to retain the materials, allowing the contractor the credited amount, if any. This is probably best catered for in the bill by including two cash columns against the items, one being headed 'credits'.

Shoring and scaffolding incidental to demolition, repair and alteration work are deemed to be included together with making good any damage caused by the erection or removal. However, shoring of structures that are not being demolished is measured as an item. Any temporary roofs and screens required are measured as metre square stating details of weatherproofing and dustproofing if required. Details must be given of any toxic or special waste likely to be encountered during the work.

Preambles

Inclusion of special preambles in the bill for this type of work is necessary in order to advise the contractor of any restrictions or other matters such as phasing of the work. Furthermore, requirements common to several descriptions can be included as preambles in order to reduce the length of the descriptions: for example, 'cutting openings is to include for quoining up jambs with brickwork toothed and bonded in gauged mortar to the existing work'.

Demolitions

Demolitions are divided into three main categories:

- Demolishing all structures
- Demolishing individual structures
- Demolishing parts of structures (excluding cutting openings, removing finishings etc.).

All demolition work is measured as items, and sufficient information must be given in the description to enable identification and the levels down to which the demolition is to be carried out. If any particular methods of demolition are required or any restrictions are to be imposed, for example on the use of explosives, then these have to be mentioned. Particulars have to be given of any services to be diverted, maintained or sealed off temporarily. Making good the parts of the structure remaining and any materials to be stored for reuse or retained by the employer has to be detailed. Breaking up parts of the building below the ground is usually left to be included as extra over the excavation items in the measured work.

Temporary supports of the remaining structure, adjoining structures, roads and other features are measured as items providing a description of what is to be shored up, details of the type of shoring required and the length of the exposed edge to be supported, with its average height.

Temporary works such as roofs, screens, floors and roads are to be measured in square metres detailing if they need to be dustproof, waterproof, weatherproof or fireproof.

Decontamination is measured as items covering the removal of hazardous materials, decontamination of remaining structures and infestation removal, with details being provided of the scope of work, the type of decontamination and whether it is to be carried out prior to demolition or during demolition activities.

Recycling is measured as an item, giving details of what is to be recycled and any limitations imposed by the local authority or recycling body.

Alterations, Repairs and Conservation

All works of alteration are measured as items and described, giving details sufficient to describe the location and extent of the work. Care must be taken to ensure that the item can be identified and that details are given of any required making good of the structure or finish. If any new work is included, then its description must be equal to that required in NRM2 for the measurement of new similar work.

Removing work is measured in an appropriate unit depending on what is to be removed, with any disconnection and/or reconnection being included in the description.

Cutting or forming of openings or recesses and filling of openings or recesses may be measured as a superficial, linear or enumerated item as appropriate, each with a dimensioned description. The details of new work should be equivalent to the information required by the NRM2 rules for similar work. The re-use of any materials should be stated, for example:

- Removing existing structures and replacing them
- Preparing existing structures for connection or attachment of new work Repairing.

Repointing joints is measured lineally, describing the nature of the joint and the type of material to be used. The width and depth of the joint need to be stated, and the removal of existing material and preparation of the surfaces deemed included.

Repointing generally can be measured in square metres, lineally or by number at the discretion of the surveyor depending on the work to be measured and described. Information should be given in the description of such matters as size and depth of raking out, type and mix of pointing, bond and size of bricks and so on.

Rules are also included in the SMM for resin or cement impregnation and injection, removing stains, cleaning surfaces, inserting wall ties, re-dressing to form new profiles and artificial weathering.

Renovating and conserving can be measured in square metres, lineally or by number at the discretion of the surveyor to reflect the size and extent of the work, with details provided of the nature of the renovation or conservation required, and the materials and methods to be used.

Decontamination, temporary works and roads and recycling for this section are measured as in the 'Demolitions' section.

Example 17

Spot Items

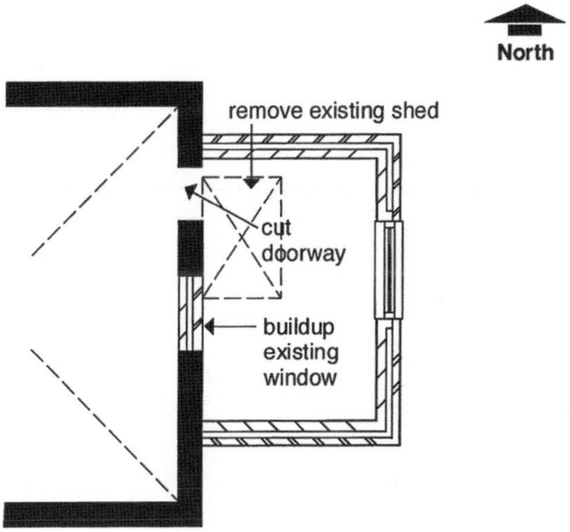

PLAN **Scale 1:100**

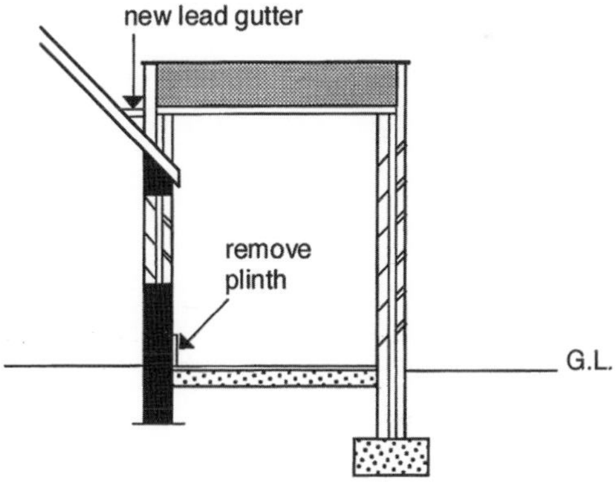

SECTION **Scale 1:100**

Fig. 43

Example 17

Taking-off list	NRM2 reference
Take down buildings	3.1.2.1
Break up floor	3.1.3.1
Filling openings	4.6.1
Forming openings	4.3.1
Concrete lintel	13.1.1
Cut back plinth	4.5.1
Work to eaves	4.2.2
Roof covering adjustment	4.9.1

				Spot Items 1
	Item		Pull down to ground level existing shed adjacent to east wall of kitchen approximately 1 900 x 1 300 and 2 000 mm high; include making good remaining brickwork and facings	3.1.2.1
	1.90 1.30		Break up and clear away concrete floor and hardcore under (assume 150 mm concrete and 150 mm hardcore).	5.7.2.3 An assumed thickness needs to be taken unless existing drawings are available. If, subsequently, this is not correct, then an adjustment to the price may be necessary.
			ALTERNATIVELY	
	Item		Break up and clear away concrete floor and hardcore under approximately 2.5 m^2 (assume 150 mm concrete and 150 mm hardcore).	3.1.3.1

				Spot Items 2
	Item		Works of alteration: take out and clear away window 1 200 x 1 200 mm in east wall of kitchen complete with frame and stone cill under. Fill opening with skin of half brick wall in commons and skin of 100 mm insulating blocks with cavity and ties' width to suit existing wall thickness. Cutting and bonding to jambs and pinning up to soffit. Plaster both faces, and adapt and extend wall tiling on one side.	3.1.2.1 The making good of existing items is deemed included. Alternatively, the window could have been removed under 4.2.1.1 and then the filling of the opening taken under 4.6.1.2 measured in square metres.
			Form opening	
	Item		Cut opening 850 x 2 050 mm in 1 ½ brick wall adjoining last window for new door opening, cut out and insert new lintel (measured separately) face up jambs, and make out flooring and skirting to new opening.	4.3.1
			Door 750 frame 2/50 100 bearing 2/150 300 1 150	
	1		327 x 140 precast concrete lintel 1 150 mm long reinforced with and including No. 3 12 mm mild steel bars	13.1.1

				Spot Items 3
			New door to take	A reminder to ensure this work is not missed
	Item		Cut back projecting brick plinth 25 x 300 mm on the east elevation for a length of approximately 4 400 mm.	4.5.1
	Item		Removing eaves gutter and fascia for a length of approximately 4 400 mm, including providing stop ends to existing gutter	4.2.2
	Item		Removing roof coverings and cutting back projecting rafters for a length of approximately 4 400 mm; prepare 1 ½ brick wall for raising.	4.9.1
			Gutter and eaves work to take	Reminder as before

Chapter 20
Preliminaries and Other Priced Bill Sections

Generally

The drafting of the preliminaries, risks, provisional sums, works to be carried out by statutory undertakers and dayworks sections of the bill, which are not usually derived from the dimensions, is discussed in this chapter. These sections are often left to a late stage in the bill preparation, as reference to the draft bill of measured items may be necessary. The specification will form the basis of these sections; if items have been lined through in the specification as they are measured, then those remaining may have to be either included or, if the specification is being provided with the tender documents, referred to in these parts of the bill. It should be noted, however, that specification items that relate to method and quality of workmanship are usually excluded from the bill unless they affect cost.

Often, companies will use standard preliminaries as the basis of creating the preliminaries bill to reduce drafting time and force the quantity surveyor to continually question the need for particular items. However, care needs to be taken to ensure that specific items required for the project are not overlooked.

Preliminaries and general conditions section

It must be remembered that the bill of quantities has to set out all the circumstances and conditions that may affect the contractor's tender. Items that are of a general nature and do not necessarily relate to the quantity of permanent work are set out in the preliminaries section for pricing by the estimator. Work Section 1 of NRM2 comprises the rules for describing

Willis's Elements of Quantity Surveying, Twelfth Edition. Sandra Lee, William Trench and Andrew Willis.

Published 2014 by John Wiley & Sons, Ltd.

and quantifying preliminaries. It is not possible to quantify the main contractor's preliminaries in the way that a work section is quantified as it is the contractor's responsibility to interpret the information provided in the tender documents and to make necessary allowances for his or her chosen method of working and the resources required to complete the project. For this reason, the preliminaries bill usually looks more like a list of headings to be priced. The section is divided as follows:

- Part A: Information and requirements (i.e. dealing with the descriptive part of the preliminaries)
- Part B: Pricing schedule (i.e. providing the basis of a pricing document for preliminaries).

Both Part A and Part B are split to cater for whether the bill is for the main contract works or a works package. There are differences between these sections; however, they cover similar areas, and for clarity only the main works portion is covered here.

Part A

Project particulars (NRM2 1.1)
This section sets the scene as to the proposed contract and full particulars of the project, including the location and details of the employer and consultants.

Drawings (NRM2 1.2)
The bill should contain a list of drawings from which the bill was prepared and a list of drawings and other documents related to the contract but not included in the tender documents. There should also be a statement explaining how the preconstruction information is dealt with in the tender documents.

At the commencement of each work section in NRM2, the exact nature of the drawn information to be given is set down. Three types of drawings have to accompany the bill of quantities:

- Block plan – to identify the site and outline of the building in relation to the town plan or other wider context
- Site plan – to locate the position of the building in relation to setting out points, means of access and general layout of the site
- Plans, sections and elevations – to show the position occupied by the various spaces in a building and the general construction and location of the principal elements.

The site and existing buildings (NRM2 1.3)
A full description of the site and details of any existing buildings should be given if their presence is likely to affect the cost of the new work. Examples are a mediaeval listed building or a multi-storey car park with continuous traffic movements.

Description of the works (NRM2 1.4)
Details and dimensions of the new buildings and how they are to be constructed must be set out by the provision of either drawn information or measurement descriptions.

Contract particulars (NRM2 1.5)
Particulars of the contract have to be given, but if a standard published form of contract is being used, then the conditions need not be repeated in full in the bill provided that the form of contract together with any relevant amendments is named and the numbers and titles of the clauses are listed. The decision as to which clauses require pricing is left to the estimator. If the conditions of contract are not standard, then the conditions must be included in full in the bill or alternatively presented as a separate volume with the tender documents and listed in the bill for pricing as shown in this chapter.

Special clauses may sometimes be added to the conditions of a standard form of contract, and these together with any modifications to clauses should be detailed in the bill. Such amendments must also be made to the actual contract that is to be signed by the employer and successful contractor, as it is not advisable to rely on a clause in the bill alone even though it may be one of the contract documents.

Employer's requirements (NRM2 1.6–1.12)
This section provides considerable detail on the provision, content and use of documents provided as part of the tender documents and the management of the works. This is followed by requirements for quality standards and quality control; security, safety and protection; specific limitations on method, sequence and timing; site accommodation, services, facilities and temporary works; and operation and maintenance of the finished building.

Part B

Employer's requirements and main contractor's cost items (NRM2)
In Part B, the pricing schedule tables list the components to be priced and identify what is to be included in the detailed description. The pricing tables give the unit of measure to be used for each item and

require the items to be classified as either a 'fixed charge' or a 'time-related charge'.

A *time-related charge* is one that is considered to be proportional not to the quantity of the item but to the length of time taken to execute the work. A *fixed charge* is one that is considered to be proportional to neither the quantity of work nor the time taken.

Broadly speaking, this part of the preliminaries section contains two types of items: firstly, those that have a separate identifiable cost such as insurance, site facilities and various fees payable, and, secondly, those costs that arise from a particular method of carrying out the work. The latter would include fixed costs for items such as providing the plant and bringing it to and removing it from the site, and time-related costs for items such as maintaining the plant on-site or for providing supervision for the works. Employers' requirements, which have to be described as fixed and time related, include the provision of site offices, fences and name boards, insurances and other fees and charges. Limitations such as working hours and sequence of work required by the employer have also to be given. Contractors' general cost items, which again have to be described as fixed or time related, include for example supervisory staff; site accommodation; facilities such as water, power and lighting; and plant and temporary works.

Risks (NRM2 Part 2–2.10)

At the time of preparing a bill of quantities, there will normally still be a number of risks remaining to be managed by the employer and the project team. It may be considered appropriate for some of these to be managed by the contractor. A schedule of construction risks should be included in the bill of quantities for those risks that the employer wishes to transfer to the contractor.

The price inserted against these risks (effectively, the contractor's estimation of the time and cost implications should the risk arise) is to be paid, irrespective of whether the risk occurs.

In order for the contractor to effectively price the risks, they should be fully described so that it is transparent what risk the contractor is required to manage, and precisely what the employer is paying for. The allowances for construction risks do not include an amount for overheads and profit. These are included in a separate section.

Provisional sums (NRM2 Part 2–2.9.1)

These are included for work for which there is insufficient information available for proper description. Provisional sums may also be included to cover possible expenditure on items that may be required but for which there is no information available at tender stage. Here, NRM2 states that provisional sums may be for either undefined or defined work (i.e. work that can be described fairly fully). For the latter, the contractor is expected to have taken the work into account when pricing other sections, such as indirect costs in preliminaries. The effect of this is that when the actual sum expended is ascertained for the final account, no adjustment to other prices in the bill is made. On the other hand, if the work is undefined, then other prices such as, for example, plant items may have to be adjusted when the work is valued for the final account.

All provisional sums shall be exclusive of overheads and profit.

Works by Statutory Undertakings (NRM2 Part 2–2.9.4)

Works that are required to be carried out by a statutory undertaker are included as a 'provisional sum' under this separate section. The scope of the works is described, and the contractor is deemed to have made allowance for any time and cost implications of the works, including all general attendances.

The provisional sums for work to be carried out by statutory undertakers do not include an amount for overheads and profit. The bill includes a separate section for this.

Daywork (NRM2 Part 2–2.13.3)

Daywork is work for which the contractor is paid on the basis of the cost of labour, materials and plant plus an agreed percentage for overheads and profit. Payment in this way is usually reserved for items that cannot be measured and priced in the normal way. Daywork payments may arise in contract variations for items such as breaking up unexpected obstructions in the excavations or for adjustment of provisional sums. To enable a percentage rate for overheads and profit to be established at the tender stage, it is necessary to include items for these in relation to dayworks in the bill and, in order to obtain competitive rates, these items should be included in such a way that they affect the total amount of the tender.

One method of achieving this is to include a number of hours for both labourers and craftsmen which the estimator prices at an hourly rate and extends into the cash columns of the bill. Some surveyors split these hours into various crafts; of course, it needs to be made clear that these rates must include for all charges in connection with the employment of labour, plus profit and overheads and so on. Sums can also be included in the bill for plant and materials; the tenderer is invited to add a percentage that represents what will be added to the net cost of these items for incidental costs, overheads and profit, should the need for daywork arise.

Chapter 21
Bill Preparation

As described in Chapter 5, the widespread utilisation of computerised systems has to a great extent made the labour-intensive manual processing of dimensions redundant, and therefore only brief mention of such processes is made here. Preparation of the bill itself is addressed in more detail because (although it too is now often part of an automatic process), as with setting down dimensions, it is important to understand the process involved in structuring and coding of the bill; thus, when computerised systems are used, the surveyor can ensure that the bill is complete and correct and that the necessary level of information is being provided to the contractor for pricing.

Abstracting

In splitting up the building into its constituent parts for measurement, the taker-off follows a systematic method, but this then creates a repetition of the same description in different parts of the dimensions. For example, facing brickwork will be measured when the external wall is measured, and again it will be in the windows and doors element as it is deducted during the measure of the openings.

Under the traditional method of processing dimensions, the function of the abstract was to collect similar items together and to classify them into sections and subsequently, according to certain accepted rules of order and arrangement, to put them in a suitable sequence for writing the bill. A manual abstract was prepared by copying the descriptions and squared dimensions from the taking-off onto abstract paper in a tabulated form as nearly as practicable in the order of the bill.

Willis's Elements of Quantity Surveying, Twelfth Edition. Sandra Lee, William Trench and Andrew Willis.

Published 2014 by John Wiley & Sons, Ltd.

The need for an abstract is eliminated by the use of a computerised system; however, the principles behind the abstract are retained, and a full breakdown of any item can be obtained.

Procedure

Traditionally, the writing of the bill (*billing*) involved, in theory, copying out the descriptions and quantities from the abstract in the form of a schedule or list on paper ruled with cash columns for pricing, but in practice it was a good deal more than this. The taker-off may describe an item briefly if it is in common use and leave it to the billing stage for the full description to be compiled. In fact, the term *working-up* originated from the task of working-up or expanding the brief taking-off descriptions into proper bill items. With the use of standard libraries of descriptions, as long as the correct description has been chosen, then this should not be required. However, there is a need for a final read-through of the bill by someone who understands the project details, including its specification, in order to provide a final edit and make sure that the correct items have been included.

Division into sections

The structure of bills of quantities varies depending on the complexity of the project and work measured. In certain countries it is customary to invite separate tenders for each trade, but in the United Kingdom, unless a form of management contracting has been adopted, it is usual to invite tenders from a general contractor only. Where separate tenders are invited, the need for separate bills for each trade is obvious, but even where tenders are obtained from a general contractor the division into work packages is of assistance in pricing and simplifies the estimator's task in respect of the parts of the work to be sublet. Moreover, it is this first step in the subdivision of the bill which enables items to be easily traced. The sections into which the bill is divided generally align with sub-contractors' work, and any departures, such as precast concrete bollards being contained in furniture or equipment, are soon recognised.

Work to existing buildings, work outside the curtilage of the site and work to be subsequently removed have to be billed in separate sections. Traditionally, substructure or work up to and including damp-proof course and external works are also kept in separate sections. These indications of location should assist the estimator in pricing and furthermore facilitate the valuation of, and measurement of possible amendments to, this work at a later stage. When a project consists of several buildings, it is often desirable to provide a separate section in the bill for each one.

Structure of bills

There are three principal breakdown structures for bills. They are:

(1) *Elemental*: Measurement and description are by group elements, following the arrangement for elemental cost planning as defined in NRM1. Each group element forms a separate section of the bill and is sub-divided through the use of elements, which are further sub-divided by sub-elements. (See Table 3 in Chapter 3.)
(2) *Work section*: Measurement and description are divided into the work sections defined in NRM2. (See Chapter 7 on taking-off.)
(3) *Work package*: Measurement and description are divided into employer or contractor-defined work packages. Works packages can be based on either a specific trade package or a single package comprising a number of different trades.

For certain reasons, it may be desirable to produce an elemental bill, in which the main divisions are design elements or constituent parts of the building (e.g. foundations or floor construction). The main purpose of such a bill is to assist a standardised system of cost analysis, which may be adopted particularly where buildings of a similar nature are to be repeated. Whilst in theory this type of bill should make estimating more accurate because the items are related to a particular part of the building, contractors who sublet work often have difficulty in collecting appropriate items together. Furthermore, estimators may find similar items occurring in different elements and have the additional problem of relating the prices for these separated items. To overcome these difficulties, some surveyors offer tenderers the bill in either traditional or elemental formats; computer sortation makes this particularly easy. If a bill has been coded correctly at take-off, then, by a simple computer sorting process, a trade bill priced by a tenderer can be easily turned into an elemental bill and compared with the pre-tender estimate, thus aiding cost management.

General principles

It must be remembered that in most cases the bill is a contract document, and therefore the descriptions must be absolutely clear and usually without abbreviations, other than those in common use. The biller must have sufficient knowledge of construction to understand the descriptions; if any appear to be vague or ambiguous, the taker-off should be consulted as to the exact meaning and an amendment made if necessary.

Order of items in the bill

The order of items in the bill is a matter for personal preference, but the Appendix to NRM2 provides guidance on alternatives. However, the following general principles are normally applied:

(1) Work sections as contained in NRM2, although locational sections such as substructure or external works may be required
(2) (a) Sub-divisions of work sections as contained in NRM2
    (b) Sub-divisions of different types of materials such as different mixes of concrete or different types of brick
(3) Within each subdivision in stage (2), the order of cubic, square, linear and enumerated items
(4) Labour-only items should precede labour and material items within the sub-divisions in (3).
(5) Items within each sub-division in (3) and (4) are placed in order of value, often with the least expensive item first.
(6) Preambles and PC and provisional sums usually form a separate bill, although in certain circumstances they may be contained in the appropriate work section.

Format of the bill

The bill for each work section should be commenced on a new sheet. The ruling of the paper and a typical heading are:

<u>BILL 4 SUPERSTRUCTURE</u>

MASONRY

					£	
	<u>BRICK WALLING</u>					
	<u>Common Brickwork in cement, lime and sand mortar (1:1:6)</u>					
A	Walls, half brick, vertical	97	m^2			

The first (left-hand) column is for the item number or letter references and the binding margin. Sometimes, an additional line is ruled to separate the binding margin from the reference column. The main wide column is for headings, subheadings and descriptions. Frequently, as illustrated here, an unruled portion is left at the top of each page for main headings such as work sections. The next columns contain the quantity of the item and the unit of measurement. Some surveyors prefer to enter the unit of measurement first so that the quantity is adjacent to the rate in the next column to facilitate the extension of the cash sum; others prefer to see the measurement figure separated from the cash figure so that there can be no confusion between the two. The last three columns are left blank for use by the estimator, who enters the rate for the item per unit of measurement and the cash extension. Sometimes, however, a cash amount is entered by the surveyor in the last two columns, for example when entering provisional sums.

Referencing items

It is essential that every item in the bill can be referred to and found easily. Some surveyors number each item consecutively through the whole bill, whilst others prefer to use the page number followed by a letter entered alphabetically against each item on the page. The letters 'I' and 'O' are usually omitted to avoid any possible confusion with numerals. If the number of items on a page exceeds the number of letters in the alphabet, then the lettering continues at this stage by using 'AA', 'AB', 'AC' and so on. The latter method has the disadvantage that it cannot be done finally until the bill is typed, as the draft will have a different number of items per page. On the other hand, serial numbering has the disadvantage that items inserted or deleted at the last moment disrupt the numbering sequence; the addition of a letter to the number of an inserted item overcomes this problem – or an item deleted may have 'not used' entered against it.

Units of measurement

Great care must be taken, particularly when there is a change, to enter the correct unit of measurement. For example, an item measured as a cubic item and indicated as a superficial item in the bill may result in a considerable difference in price. Even if the error is detected and queried by a tenderer, it may involve the issue of an addendum to the bill – which should be avoided if possible. To assist in avoiding mistakes, particularly when items are inserted, it is preferable to enter the unit of measurement against each item rather than using two dots to indicate repetition.

Order of sizes

Sizes or dimensions in descriptions should always be in the following order: length, width, height. Sometimes the width or front-to-back dimension of, say, a cupboard is referred to as its 'depth'. If there is likely to be any doubt, the dimensions should be identified. For example:

sink base unit 1000 mm long × 600 mm wide × 900 mm high overall…

Use of headings

Generous use of headings will not only help estimators to find their way about the bill but also may reduce the length of descriptions. Headings generally fall into one of four categories:

(1) Work section headings, such as 'Masonry'
(2) Subsection headings, such as 'Brick walling'
(3) Headings that partly describe a group of items to follow, such as 'Common brickwork in cement mortar (1:3)'.

Work section headings are often repeated in the top right-hand corner of each page. Other headings should be repeated when a new page is started. This enables it to be seen at once what is being dealt with, thus avoiding the need to search back a few pages. A new heading usually terminates a previous heading; where a new heading is not used, or if there is likely to be any doubt, 'end of …' should be entered in the appropriate place in the bill.

Writing short

It is often convenient for pricing purposes to keep items together in the bill which would, following the normal rules, be separated. For example, it is probably easier for an estimator to price the fittings on a gutter whilst dealing with the particular gutter although this means that enumerated items have to be dealt with in the middle of linear items. The method of entering these items is known as writing short; it avoids breaking completely the sequence of linear items, as follows:

D	112 mm UPVC straight half round gutter fixed to timber with standard brackets	125	m			
E	Extra for stop end	9	nr			
F	Extra for angle	7	nr			

As the example shows, written short items are inset so that the main items stand out. In the written short items, the words 'extra for' have been repeated instead of using 'ditto', thus avoiding any possible confusion with the use of 'ditto' in the main item. Items that are written short automatically refer to the main item under which they are written, so there is no need to refer back. NRM2 may require items that are often written short to be described as 'extra over'; this means that the main item has been measured over the written short item. In other words, the estimator is to allow for the extra cost of the item only as it displaces the main item.

Unit of billing

The general unit of billing for other than enumerated items is the metre. The dimensions and their collections having been taken to two decimal places, the bill entries are rounded up or down to the nearest metre. The main exceptions are steel bar reinforcement and structural steel, which are billed in tonnes to two decimal places.

Framing of descriptions

The framing of descriptions was introduced in Chapter 4. Care must be taken to leave no doubt as to their meaning, particularly as the bill is a legal document. The opening phrase of a description should indicate the principal part of the item so that the estimator knows immediately what it is about.

The word 'approved' should be avoided if it leaves any doubt about the quality of the item. Its sole justification is to point out to the contractor that the architect's approval should be obtained for the article to be used where there are several alternatives all costing about the same. If some particular material is described and the words 'or other approved' are added to the description, a gambling element is at once introduced which is directly contrary to the purpose of the bill of quantities. One estimator may decide to price for a less expensive alternative in the hope that the architect will approve it, whilst another may price the material specified. The words 'or other equal and approved' are sometimes used by public authorities after the specification of a proprietary article to avoid the criticism that a monopoly situation has been created, favouring a particular firm.

Long-winded descriptions should also be avoided, and any superfluous words omitted. The surveyor must cultivate the difficult art of making descriptions concise and easily understood and at the same time omitting nothing that is essential to the estimator for pricing. In the following, for example, the words underlined could be left out:

<u>Small</u> 19 × 19 mm wrot softwood cover fillet <u>planted on around bathroom door frame</u>
50 × 175 mm sawn softwood in floor joists <u>spiked to timber wall plates</u>.

It should be noted that in the above, if fixing is not specified, it is at the discretion of the contractor. In these cases, planting or spiking would normally be omitted as it is the most economical way of fixing. Furthermore, the location of an item is not normally required.

Consistency in spelling is important if only to make the bill look like a workmanlike document. Often, there are alternatives for spelling technical words such as 'lintel' or 'lintol' and 'cill' or 'sill'. An easy solution to this problem is to establish a *house rule* that spelling should be as shown in NRM2. Furthermore, consistency in language is important, as minor differences in phraseology may cause an estimator trouble in deciding if the meaning is intended to be different. For example:

(a) Raking cutting *to* existing Code 4 milled lead
(b) Curved cutting *on* existing Code 4 milled lead.

In this case, there is no actual difference to be inferred from the word 'on'.

Totalling pages

There are two ways in which the surveyor may indicate how the cash totals on each page are to be dealt with. Firstly, the total may be carried over to be added to the next page and so on to the end of the section. The foot of the page is completed as follows:

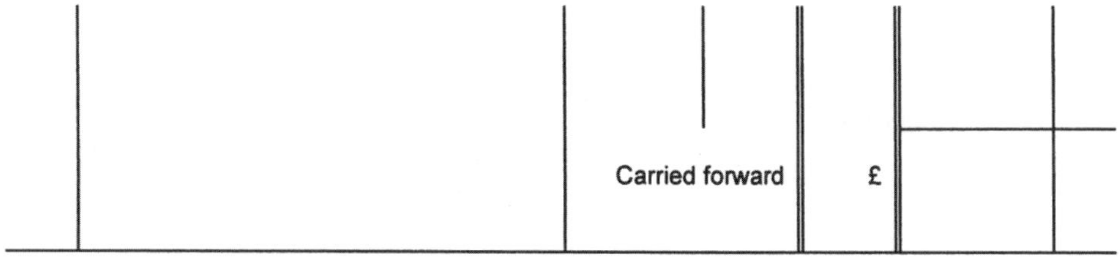

and the top of the following page as:

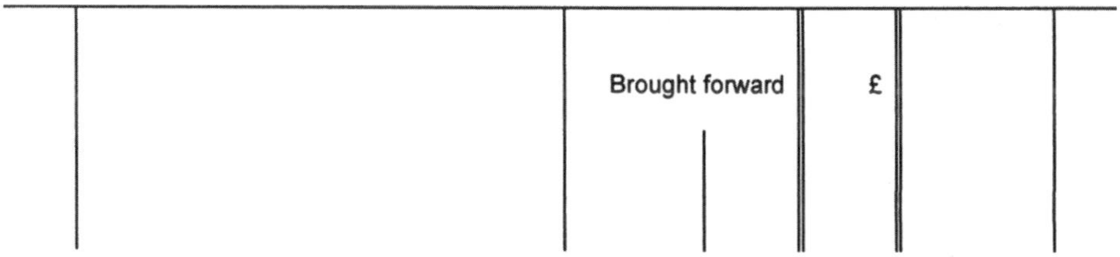

The end of the section is completed as follows:

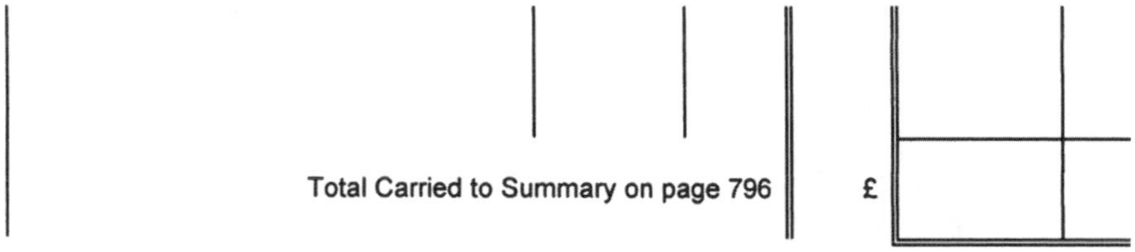

This method is suitable only when the section contains a few pages. Any mathematical error made at an early stage is carried through all the pages and, when discovered, involves several corrections and re-totalling of the pages. Generally, this method should be avoided.

Secondly, each bill page may be totalled and each of these totals carried to a collection at the end of the section. The bottom of each page is completed as follows:

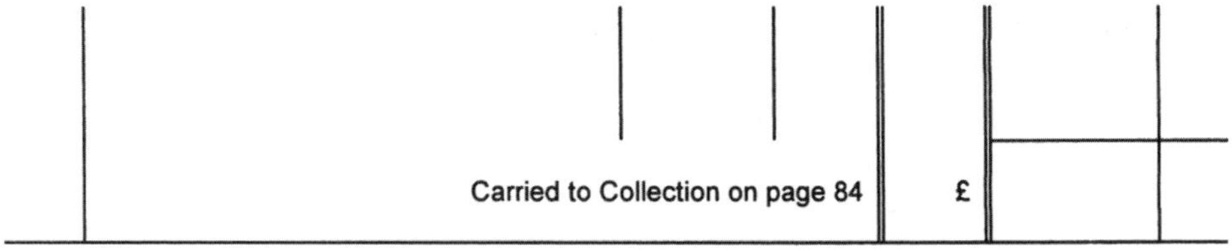

The collection is on the last page of the section as follows:

				£
COLLECTION				
Total from	page 80			
	page 81			
	page 82			
	page 83			
	page 84			
Total carried to Summary on page 796			£	

Summary

At the end of the bill, a summary is prepared for insertion of the totals from the collections of each section. To enable the total cost of any section to be established easily, it may be advisable to enter the title of each section as follows:

					£
SUMMARY					
Preliminaries	Total from	page 10			
Specification preambles		page 56			
Substructure		page 76			
etc etc					
TOTAL CARRIED TO FORM OF TENDER				£	

Sometimes a summary is made at the end of the work sections comprising the superstructure, and the total of this is carried to a general summary at the end of the bill, thus reducing the length of the latter. At the end of the final summary, provision may be made for the addition of items such as insurance and water for works, the price of which may be dependent on the total cost of the work. All such items are described fully in the preliminaries and a note made against them that they are to be priced in the summary.

The process of checking

When the bill has been produced, it must be checked very carefully. There are varying customs adopted to show that a page has been checked and by whom. The points to be looked for are as follows:

(1) In each item:
- (a) Correctness of figures
- (b) Correctness of units of measurement, especially changes from one unit to another (e.g. from cube to square or from square to linear), properly indicated
- (c) Correctness of descriptions, especially any figures contained therein, making it quite clear whether these are in metres or millimetres.

(2) Generally:
- (a) Sections and sub-sections of the bill headed properly, and headings carried over where necessary
- (b) Order of the items
- (c) Proper provision for page totals and their carrying forward or collection.

These points may appear fairly obvious, but it is extremely easy in the rush to finish a bill, which is not unusual, for some inaccuracy to be overlooked. A fascia intended to be 40 mm may be billed as 400 mm, if there has been a data entry or typing error.

Numbering pages and items

It is important to see that the pages of the draft bill are numbered in sequence before the final print is made. When the bill is complete, one should fasten it all together and look through it, making sure that all the pages are present and in the correct sequence and that all items are referenced either by letters or by sequential numbering, the latter being best done at this stage.

General final check

If undertaken manually, the whole of the dimensions and abstract should be examined carefully to ensure that all calculations have been ticked and all items have been lined through. Any item not dealt with should be drawn to the attention of the taker-off or other person concerned. An opportunity should be provided for the lead taker-off to look through the complete, bill as intentions may have been misinterpreted and errors in descriptions may stand out which might otherwise pass unnoticed. It should be noted that taking-off is not checked unless, perhaps, a junior is measuring for the first time. Therefore, it is advisable to make a check of the main quantities in the bill by making comparisons between items or by making approximate calculations, possibly on the following lines:

- Check the total cube of excavation with the total cube of disposal of soil.
- Calculate the total floor area of the building on all floors inside the external walls, and compare this with the total area of the floor finishes and beds.
- Make an approximate measurement of the external walls, deducting all openings, and compare with the total area of walling in the bill.
- Count the number of doors and windows, and compare with the total numbers of each in the bill. Ironmongery quantities also could be compared with the totals.
- Check the number and type of sanitary appliances with those billed.
- Measure the approximate area of roof tiling or flat roofing and check with the bill.
- Make an approximate check on any items that lend themselves to this, such as length of copings and eaves gutters, number of cupboards or other fittings, area of external pavings and number of manholes.
- Compare the painting items with the corresponding items in other sections as far as possible; for example, take the total area of all wall or ceiling plaster compared with the areas of decoration on plaster.

These checks must not be expected to produce an exact comparison. Their principal purpose is to ensure that the quantities are not wildly out through some serious mistake; if the figures are reasonably near those in the bill, the checks will have served their purpose. Even if, owing to pressure of time, this check is left until the bill has been sent out to tenderers, errors found may be notified to them and thus taken into account.

The final bill still needs to be checked thoroughly when produced using a computerised billing system, as the different measurers may have chosen slightly different items from the library for the same piece of work, or rogues may have been prepared with slightly different wording resulting in inconsistency and duplication of items in the final bill.

Cover and contents

The front cover of the bill should as a minimum have the title and location of the work, the date and the surveyor's name and address. The name of the employer is also often included. A sheet should be included at the front of the bill, listing the contents; this is particularly important in the case of a large bill.

Other bill types

Other titles for bills which have a special purpose may be encountered and are mentioned here for reference purposes. *Reduction bills* are special bills prepared when the tender figure is too high and a reduction in price is obtained by altering the work in some way. The bill may contain omissions and additions to the original. *Addenda bills* contain details of work required which is additional to the original design, determined after completion of the original bill.

Specialist bills may be required to obtain tenders for specialist work, such as electrical installation, which is to form nominated subcontract work. These bills should contain the full preliminaries section of the main bill and should be accompanied by the necessary drawings.

Bill of approximate quantities

This bill, also known as a *provisional bill*, is used when there is insufficient information available to prepare an accurate bill of quantities. Although the quantities are approximate, the descriptions should be correct. The bill is used to obtain rates for items from tenderers; as the production information becomes available, a *substitution bill* can be prepared and priced using the approximate bill as a basis.

Schedule of prices or rates

For smaller contracts without a bill and where the specification and drawings are the contract documents, a schedule of descriptions, without quantities, can be prepared. The contractor is asked to enter rates against the items, which are then used for valuing variations to the design. The schedule seldom attempts to be comprehensive, and the rates are often unreliable. This method has also been used for contractor selection when drawn information is so scanty that even a bill of approximate quantities is impracticable.

Figs. 44, 45 and 46 show how the dimensions for some substructure measurement are extended and checked, and then transferred to the abstract and billed. It is important that figures can be traced from the dimensions through to the bill and that, during the final account stages, figures in a bill can be substantiated or easily amended if variations occur.

			Simple concrete foundation	Notes and waste calculations	
				L	W.
			Spread	4.100	2.000
			Foundation 0.750	6.150	4.500
			Less wall 0.215	10.250	6.500
			2) 0.535	+ 0.535	0.535
			2/ 0.268	10.785	7.035
			0.535		
	10.79 7.04	75.96	Site preparation, remove topsoil for preservation average 150 mm deep & Deposit in temp spoil heaps, average 25 m distance $\times$ 0.15 = 11.39 m^3		
			Deduct	It is permissible to deduct two items in this way, but care needs to be taken in the abstract with the change in units being deducted.	
	6.15 4.50	27.68	Both last $\times$ 0.15 = 4.15 m^3		
					Girth
	32.64 0.75 0.75	18.36	Excavation foundations, n.e. 2 m deep & Backfill and foundations with excavation material		10.250 6.500 2/16.750 33.500
				less 4/0.215	0.860
				centre line	32.640
	32.64 0.75 0.23	5.63	Concrete mix A in foundations & Deduct Filling and foundations a.b. & Add Dispose of excavation material off-site.		

Fig. 44 Taking-off

Simple concrete foundations

GROUNDWORKS

Site preparation

Remove topsoil for pres. Average 150 mm deep

add		deduct	
75.96	1	27.68	1
27.68			
48.28			
48 m^2			

Excavation

Excavation foundation n.e. 2 m deep

add		deduct
18.36	1	
18 m^3		

Filling

Filling around foundations, average thickness > 250 mm, arising from excavation

add		deduct	
18.36	1	5.63	1
5.63			
12.73			
13 m^3			

Disposal

Deposit topsoil in temporary spoil heaps average 25 m

add		deduct	
11.39	1	4.15	1
4.15			
7.24			
7 m^3			

Dispose of excavation material off-site.

add		deduct
5.63	1	
6 m^3		

Fig. 45 Abstract

	Description of Work	Quantity	Unit	Rate	£	p
	GROUNDWORK					
A	Site preparation, remove topsoil for preservation, average 150 mm deep	48	m^2			
B	Excavation, foundations, not exceeding 2 m deep	18	m^3			
C	Filling to excavations, average thickness greater than 250 mm, arising from the excavations	13	m^3			
D	Disposal of excavated material off-site	6	m^3			
E	Disposal excavated material on-site in temporary spoil heaps not exceeding 25 m from excavation	7	m^3			
	To collection			£		

Fig. 46

Appendix
Mathematical Formulae and Applied Mensuration

Formulae for areas (A) of plane figures

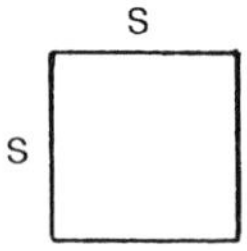

Square $A = S \times S$

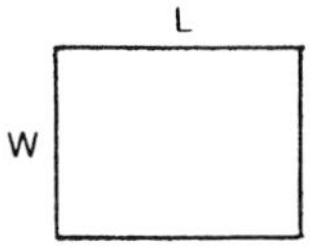

Rectangle $A = L \times W$

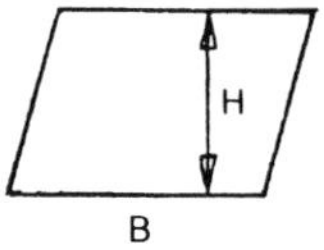

Parallelogram $A = B \times H$

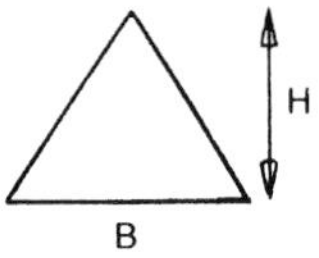

Triangle $A = \dfrac{B \times H}{2}$

Willis's Elements of Quantity Surveying, Twelfth Edition. Sandra Lee, William Trench and Andrew Willis.

Published 2014 by John Wiley & Sons, Ltd.

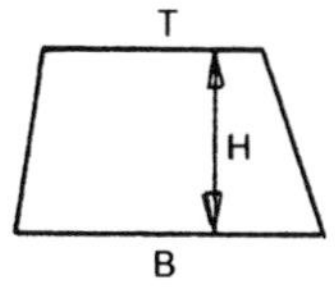

Trapezoid

$$A = \frac{(B+T)\times H}{2}$$

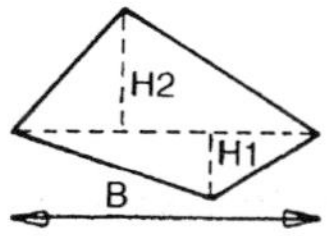

Trapezium

$$A = \frac{B\times H1 + B\times H2}{2}$$

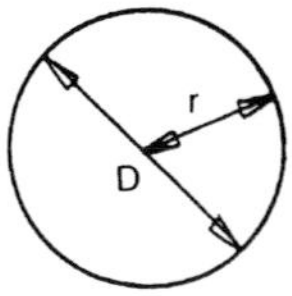

Circle

$$A = \pi \times r \times r$$

$$\text{or} \quad A = 0.7854 \times D \times D \left(\text{Note } \frac{\pi}{4} = 0.7854\right)$$

$$(\text{Circumference} = \pi \times D \text{ or } 2\pi r)$$

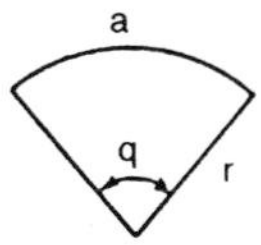

Sector of circle

$$A = \frac{r\times a}{2} \quad \text{or} \quad A = \frac{q}{360}\pi r^2$$

$$\left(\text{Note length of arc} = \text{angle}\frac{q}{360}\times 2\pi r\right)$$

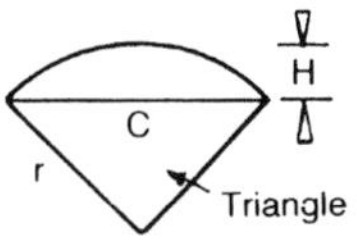

Segment of circle

$A = S - T$

Where S = area of sector

T = area of triangle or approximately

$$A = \frac{1}{2} \times \left(\frac{H \times H \times H}{C} \right) + \left(\frac{2}{3} C \times H \right)$$

Where H = rise

C = chord

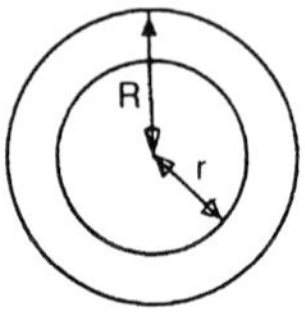

Annulus

$A = \pi(R + r) \times (R - r)$

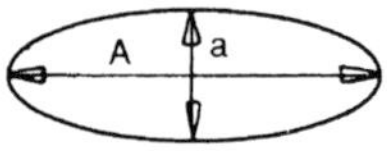

Ellipse

$A = 0.7854 \times (A \times a)$

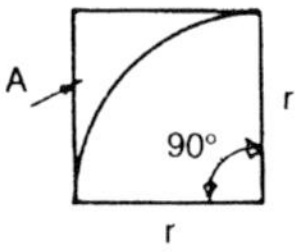

Bell mouth
(at road junction)

$A = 0.2146 \times r \times r$

Regular polygons

Pentagon	(5 sides)	$A = S \times S \times 1.720$
Hexagon	(6 sides)	$A = S \times S \times 2.598$
Heptagon	(7 sides)	$A = S \times S \times 3.634$
Octagon	(8 sides)	$A = S \times S \times 4.828$
Nonagon	(9 sides)	$A = S \times S \times 6.182$
Decagon	(10 sides)	$A = S \times S \times 7.694$

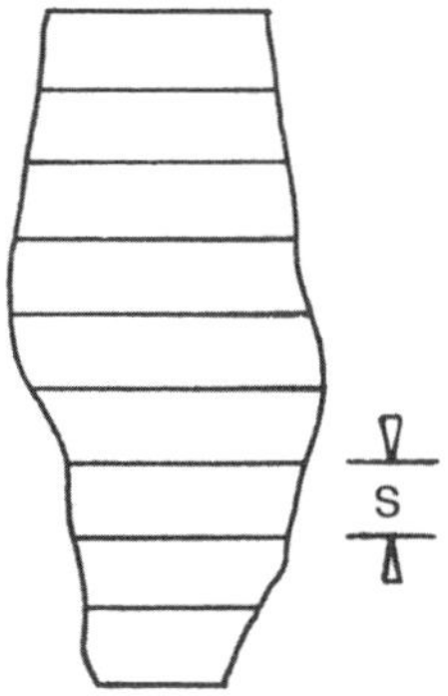

Irregular figures

Divide figure into trapezoids by equidistant parallel lines (ordinates or offsets)

$$A = S \times \left(\frac{P}{2} + Q\right)$$

(Where S = distance between ordinates
P = sum of first and last ordinates
Q = sum of intermediate ordinates)

or

Simpson's Rule (must be even number of trapezoids)

$$A = \frac{S}{3} \times \left(P + 4 \times Z + 2 \times Y\right)$$

(Where Z = sum of even intermediate ordinates
Y = sum of odd intermediate ordinates)

Formulae for surface areas (SA) and volume (V) of solids

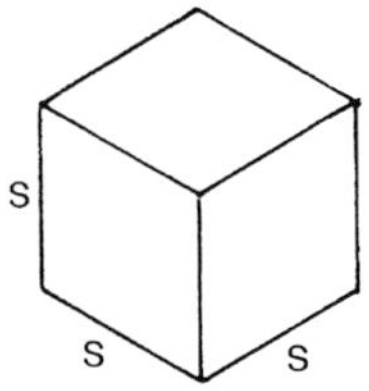

Cube

SA = 6 × S × S
V = S × S × S

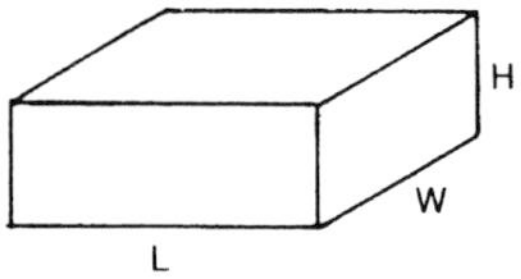

Rectangular prism

SA = 2(L × W) + 2(L × H) + 2(W × H)
V = L × W × H

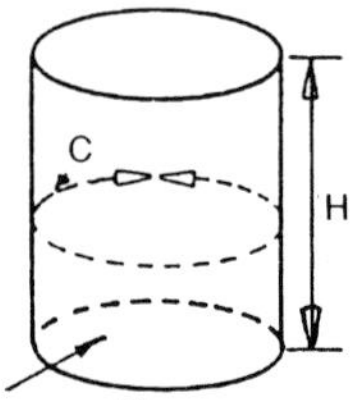

Cylinder

SA = (C × H) + (2 × B)
V = B × H
Where B = area of base (π × r × r)
C = circumference (π × D)

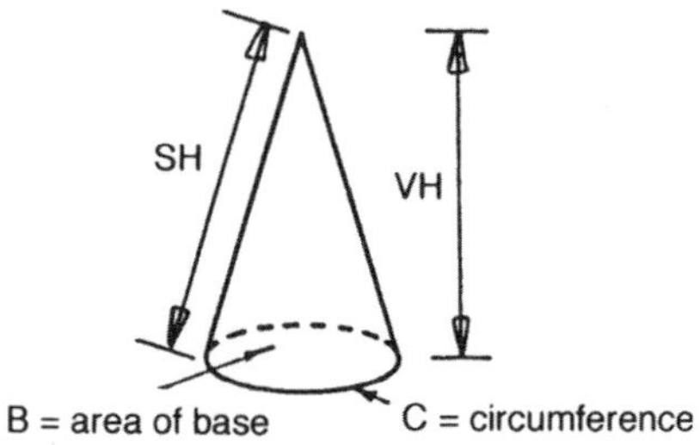

Cone

$$SA = \frac{C \times SH}{2} + B$$

$$V = \frac{B \times VH}{3}$$

Where B = area of base ($\pi \times r \times r$)
C = circumference of base
SH = slope height

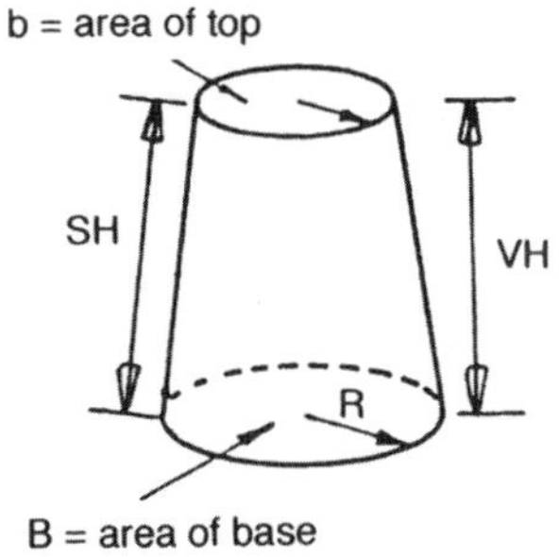

Frustum of cone

$$SA = \pi \times SH \times (r + R) + b + B$$

$$V = \frac{VH}{3}\left(\pi r^2 + \pi R^2 + \sqrt{B \times b}\right)$$

Where B = area of base
b = area of top
R = radius at base
r = radius at top
SH = slope height

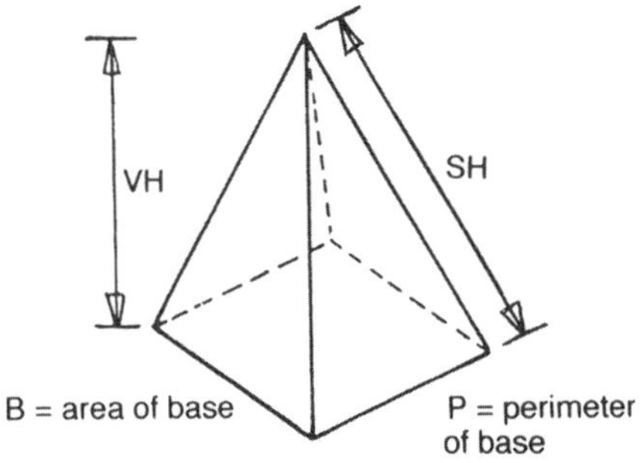

Pyramid

$$SA = \frac{P \times SH}{2} + B \text{ (regular pyramid only)}$$

$$V = \frac{B \times VH}{3}$$

Where B = area of base
P = perimeter of base
SH = slope height

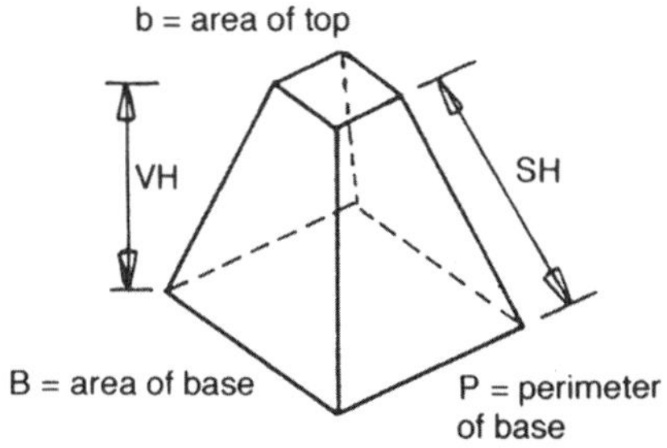

Frustum of pyramid

$$SA = \frac{SH}{2} + (p + P) + B + b \text{ (regular figure only)}$$

$$V = \frac{\left(B + b + \sqrt{B \times b}\right) \times VH}{3}$$

Where B = area of base
b = area of top
P = perimeter of base
p = perimeter of top
SH = slope height

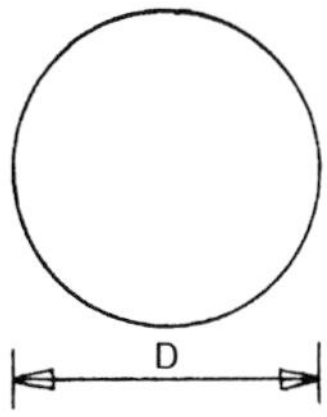

Sphere

$SA = \pi \times D \times D$

$V = 0.5236 \times D \times D \times D$

$\left(\text{Note } \frac{\pi}{6} = 0.5236\right)$

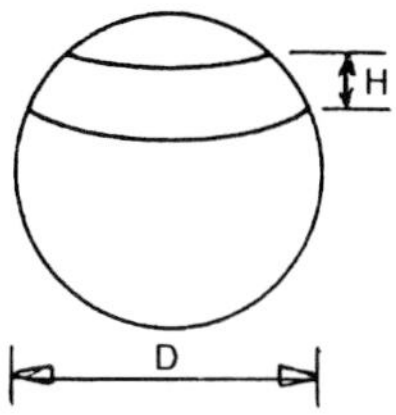

Zone of sphere

$SA = \pi \times D \times H$ (excluding base & top)

$$V = \frac{\pi \times H}{6} \times (3 \times R \times R + 3 \times r \times r + H \times H)$$

Where R = radius at base
r = radius at top

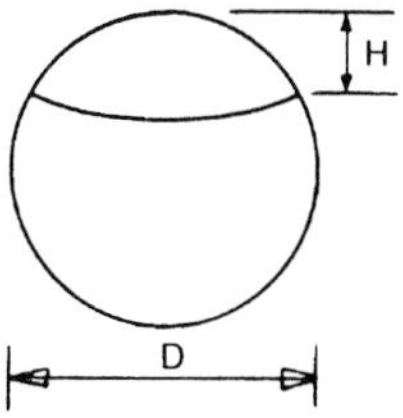

Segment of sphere

$SA = \pi \times D \times H$ (excluding base)

$$V = \frac{\pi \times H}{6} \times (3 \times R \times R + H \times H)$$

Where R = radius of base

Irregular areas

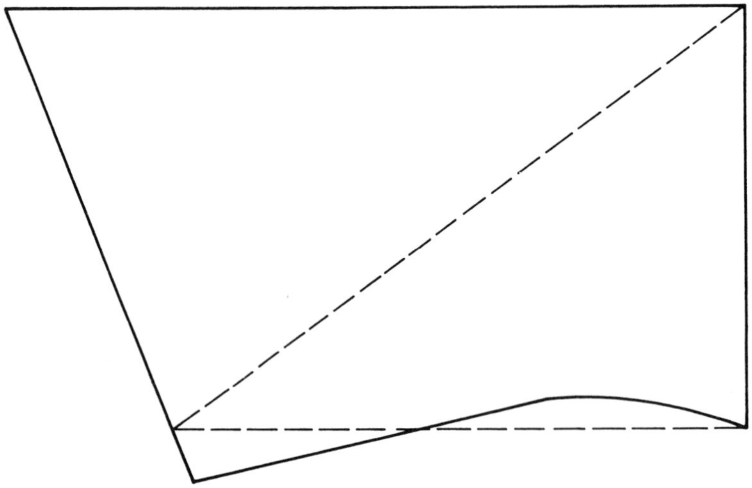

Fig. A.1

Any irregular-shaped area to be measured is usually best divided up into triangles, the triangles being measured individually and then added to give the area of the whole-if one of the sides, as for instance in the case of paving, is irregular or curved, the area can still be divided into triangles by the use of a compensating, or give and take, line, i.e. a line is drawn along the irregular or curved boundary in such a position that, so far as can be judged, the area of paving excluded by this line is equal to the area included beyond the boundary. In Fig. A.1 the area of paving to be measured is enclosed by firm lines, the method of forming two triangles (the sum of of which equals the whole area) being shown by broken lines.

For a more accurate calculation of the irregular area, particularly if evenly spaced offsets are available dividing the area into an even number of strips, Simpson's rule may be applied. The intermediate offsets should be numbered as it is necessary to distinguish the odd numbers from the even. The formula is given on page 311.

Where two sides of a four-sided figure are parallel to form a trapezoid, it is not necessary to divide the figure into triangles, as the area equals the length of the perpendicular between the parallel sides multiplied by the mean length between the diverging sides. In Fig. A.2 the area is EF × GH, GH being drawn halfway between AB and CD and being equal to

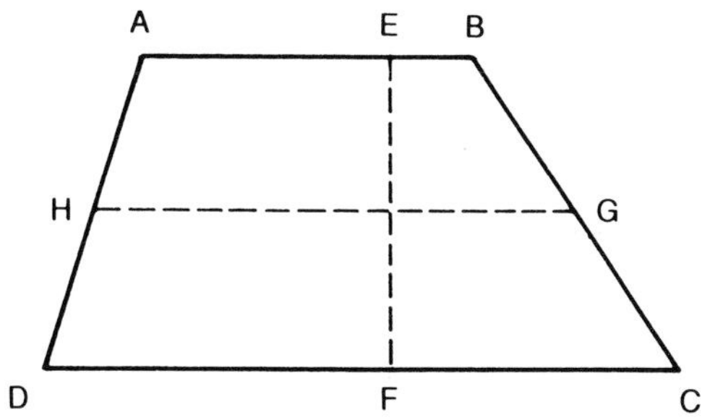

Fig. A.2

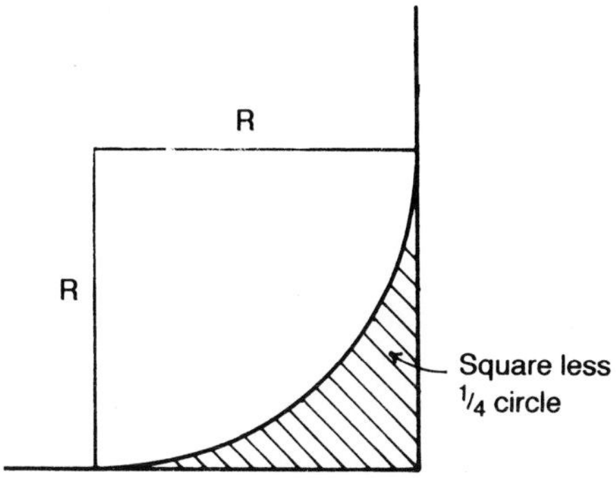

Fig. A.3

$$\frac{AB + CD}{2} \quad \text{i.e. the average of AB and CD}$$

Another irregular figure that often puzzles the beginner is the additional area to be measured where two roads meet with the corners rounded off to a quadrant or bellmouth. This is most easily calculated as a square on the radius with a quarter circle deducted. For example, consider Fig. A.3.

$$\begin{aligned}
\text{Additional area} &= R^2 - \frac{1}{4}\pi R^2 \\
&= R^2 - \frac{11}{4}\pi R^2 \\
&= \frac{3}{14} R^2
\end{aligned}$$

Excavation to banks

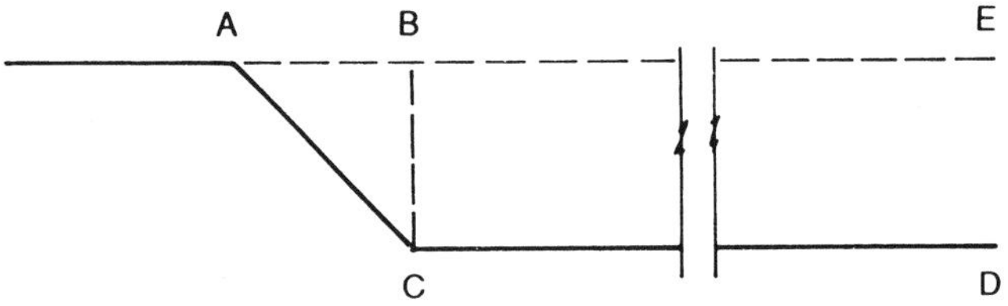

Fig. A.4

The volume of excavation necessary on a level site to leave a regular sloping bank is the sectional area of the part displaced (triangle ABC in Fig. A.4) multiplied by the length; the volume of the remaining excavation equals the sectional area of the rectangle BCDE also multiplied by the length. It may be, however, that the natural ground level is falling in the length of the bank, as shown in Fig. A.5. The volume of earth to be displaced should then theoretically be calculated by the prismoidal formula. As the final volume is required in cubic metres the calculation is carried out in metres:

$$V = \frac{L(A + a + 4m)}{6}$$

Where:
$V =$ Volume
$L =$ Length
$A =$ Sectional area at one end
$a =$ Sectional area at the other end
$m =$ Sectional area at the centre.

Note that in Fig. A.5 the area m is not the average of areas A and a, but should be calculated from the average dimensions.

If AB and BC in Fig. A.4 are both 2.00 m at the higher end and 1.00 m at the lower end, they would, assuming a regular slope, be 1.50 m at the centre. The volume of the prismoid in Fig. A.5 in cubic metres would therefore be:

$$15\left(\frac{2\times 2}{2} + \frac{1\times 1}{2} + 4\times\frac{1.5\times 1.5}{2}\right)$$

$$= \frac{15(2 + 0.50 + 4.50)}{6} = \frac{105}{6}$$

$$= 17.50\,\text{m}^3$$

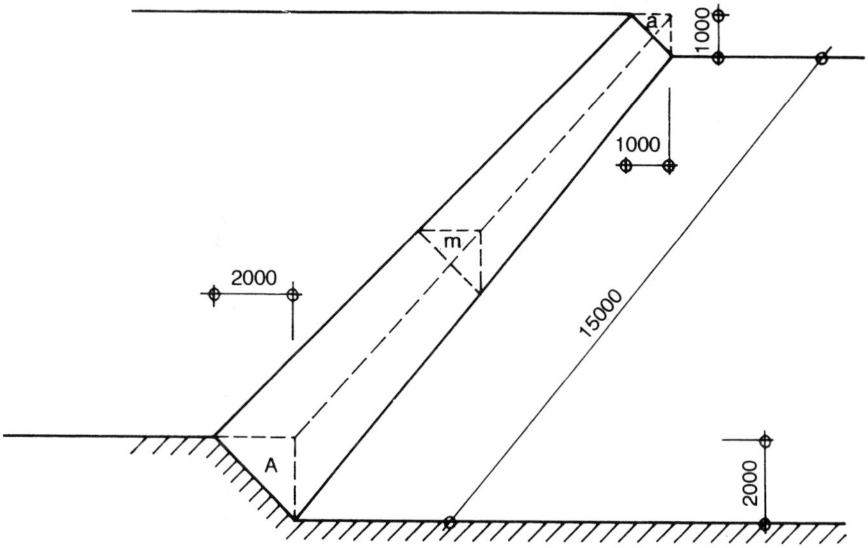

Fig. A.5

In practice, however, it will often be found that so precise a calculation is not made in such cases. The surface of ground, whether level or sloping, is not like a billiard cloth, and the natural irregularities prevent the calculation from being exact. Moreover, in dealing with normal construction sites the excavation for banks is a comparatively small proportion of the whole excavation involved (unlike, say, the case of a railway cutting); any departure from strict mathematical accuracy due to the use of less precise methods would involve a comparatively small error. In the example given above the volume might in practice be taken as the length multiplied by the sectional area at the centre, i.e.

$$15 \times \frac{1.50 \times 1.50}{2} = 16.88\,\text{m}^2$$

Although an error of 0.62 m^3 may be thought high, it must be remembered that it is being assumed that the excavation to the bank is only a small proportion of the whole.

If the natural ground level falls to such an extent that it reaches the reduced level, and the bank shown in Fig. A.5 therefore dies out to nothing (as in Fig. A.6), the formula will be found to simplify itself. a becomes zero. m becomes by simple geometry $\frac{A}{4}$, because the triangles at A and m are right-angled triangles, that at m having two sides enclosing the right angle each half the length of the corresponding sides of the triangle at A.

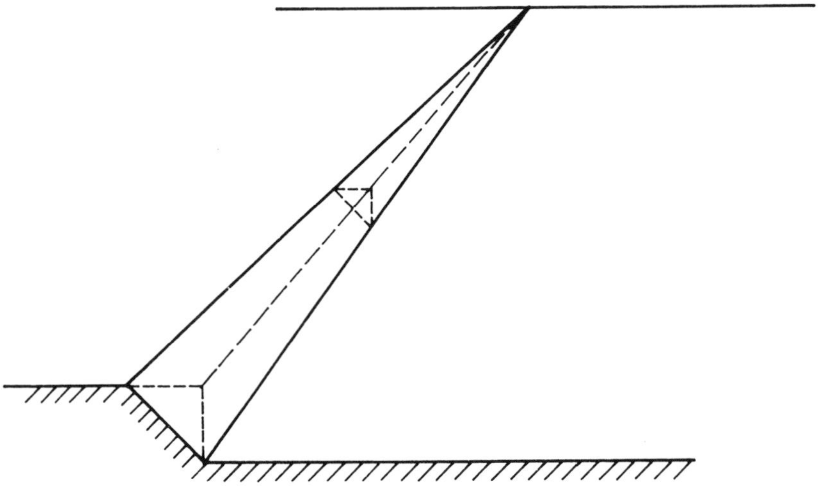

Fig. A.6

$$V = L\frac{\left(A + 0 + 4 \times \frac{A}{4}\right)}{6} = \frac{L \times A}{3}$$

which is the formula for the volume of a pyramid.

If the volume of earth to be excavated forms an even number of prismoids of equal length, then Simpson's rule may be applied taking the area at the offsets rather than the length as in the case of the irregular area in Fig. A.5.

Excavation to sloping sites

The theoretical principle for measuring the volume of excavation in cutting for a sloping bank may be extended to a sloping site (Fig. A.7) If the figures given in the diagram are natural levels and it is required to

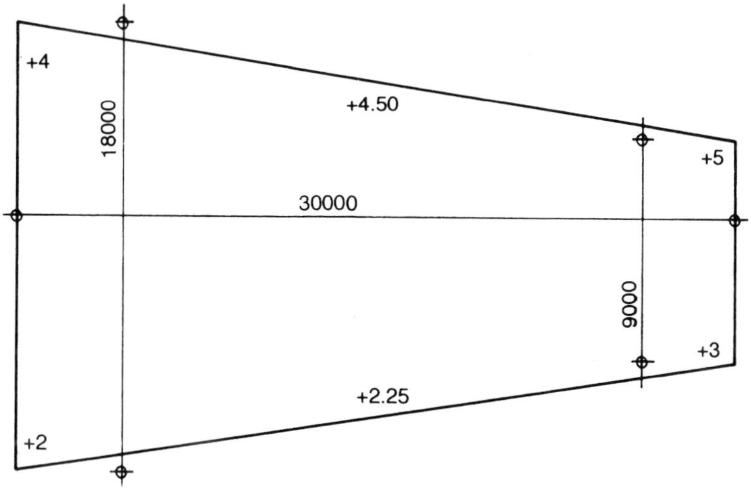

Fig. A.7

excavate to a general level of +1 (the end boundaries of the area being parallel), the volume of excavation may be calculated by the prismoidal formula already given:

$$V = \frac{L(A + a + 4m)}{6}$$

When:

$$L = 30$$

$$A = 18 \times \frac{3+1}{2} = 36$$

$$a = 9 \times \frac{4+2}{2} = 27$$

$$m = 13.50 \times \frac{3\frac{1}{2} + 1\frac{1}{4}}{2} = 13.50 \times \frac{19}{4} \times \frac{1}{2} = \frac{256.50}{8}$$

then:

$$V = \frac{30(36+27) + \frac{256.50 \times 4}{8}}{6} = \frac{30(36+27+128.25)}{6}$$

$$= 5 \times 191.25 = 956.25\,\text{m}^3$$

If only a simply average of the four corners were taken, the result would be:

	4
	5
	3
	2
÷ 4)	14
Average ground level	3.50
Reduced level	1.00
Average depth	2.50

$30 \times 13.50 \times 2.50 = 1012.5\,m^3$

This shows an error that is greater because the intermediate level on the lower boundary is not an average of the two end levels.

An average of the eight levels (giving the centre levels double value) would be:

	4
2/4.50 =	9
	5
	3
2/2.25 =	4.50
	2
÷ 8)	27.50
	3.44
	−1.00
	2.44

or a volume of $30 \times 13.50 \times 2.44 = 988.2\,m^3$, from which it will be seen that there is a definite error. This method would be satisfactory, however, if the area of the ground were a rectangle.

The error would vary with the regularity of the slope, and where the slope is fairly regular it may often be sufficiently accurate to take an average depth over the whole area. For an ordinary site, calculation would probably be made in this way in practice.

If the formula is to be applied where end boundaries are not parallel it will be necessary to draw parallel compensating lines along these two boundaries. It is, of course, assumed that slopes between the level given are regular. If a line of intermediate levels were given between the upper and lower lines in Fig. A.7, it would be necessary to treat the two portions separately.

Grids of levels

On larger sites it is customary to take a grid of levels over the whole area, or at least over the area of the proposed construction, at regular intervals forming squares, plus additional levels at any significant points such as existing manhole covers or along embankments. When calculating the average level of an area covered by a grid, it is necessary to find the average level of each square of the grid by totalling the levels at the four corners and dividing by four. The average levels thus found are added together, the total is divided by the number of squares and the result gives the average level of the area. To calculate the amount of excavation the total area is multiplied by the difference between the average level of the area and the formation level required. If the excavation of top soil has been measured as a separate item then the depth of this should be deducted from the average level.

A quicker method is to total the levels in each of the categories indicated below and multiply each total by the appropriate weighting shown.

(1)	Levels at the external corners on the boundary of the area	Multiply by 1
(2)	Levels at the boundaries of the area (other than those in (1) or (3))	Multiply by 2
(3)	Levels at any internal corners on the boundary of the area	Multiply by 3
(4)	Intermediate levels within the area	Multiply 4

The results obtained are totalled and divided by a number equal to four times the number of squares (or the total number of weightings). The result is the average level of the area and will produce the same result as the first method.

Example of weighted average excavation

Assume a grid of 4 squares with each point 5 m apart, as shown in Fg. A.8. The existing levels are included in the table and the formation that this needs to be excavated down to is at 62.000.

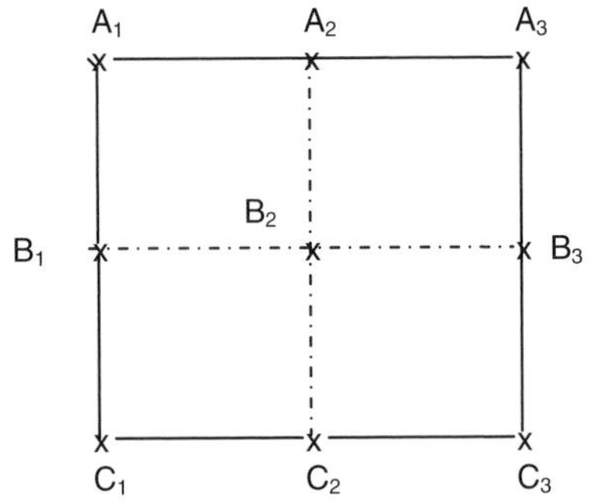

Fig. A.8

Existing ground levels

	×1		×2		×4
A_1	62.507	A2	62.750	B2	63.500
A_3	63.100	B1	62.900		
C_1	63.500	B3	63.800		
C_3	64.100	C2	63.800		
	1 × 252.950		2 × 253.250		
			506.500		254.000
					506.000
					252.950
					16)1013.450

Weighted average ground level	=	63.341
Less formation level	=	62.000
Depth of reduced level excavation	=	1.341

This would now be transferred to the dimension paper as follows:

4/	5.00 5.00 1.34		Excavate to reduce levels, maximum depth ≤ 2 m

Sometimes a site may be partly excavated and partly filled and it will be necessary to plot a cut and fill contour using interpolation, as shown below, to ascertain the location. Contours may also to be plotted to represent the division between depth bands of excavation or fill as required by the SMM. Grid squares that are cut by the contour line to form triangles or trapezia should be dealt with separately; the average level of each is found by adding the levels at the corners and dividing by the number of corners. The area of each irregular figure is multiplied by the difference between its average level and the formation level to give the volume of excavation or fill to add to the main quantity.

When measuring excavation work, one should always look for sudden changes in levels which may indicate items such as embankments or craters. These areas should be dealt with separately and not averaged within the main area as above.

Interpolation of levels

When measuring excavation work, it is sometimes necessary to ascertain the ground level at a point between two levels given on the drawing or to locate a point at which a certain level occurs. This can be achieved by interpolation as shown in the following example:

Give two levels at points A and B 50 m apart, find an intermediate level at point C 20 m from A.

Level at A = 3.60
Level at B = 2.90
Difference 0.70
Therefore the ground falls 0.70 over 50 m
To find the fall y over 20 m

$$\frac{0.70}{50} = \frac{y}{20}$$

$$20 \times 0.70 = 50y$$

$$y = 0.28$$

The level at C = 3.60 – 0.28 = 3.32

If the intermediate level is known, such as a formation level, then the distance from A could be found in a similar manner, the unknown being the distance rather than the level.

Index

Willis's Elements of Quantity Surveying, Twelfth Edition. Sandra Lee, William Trench and Andrew Willis.

Published 2014 by John Wiley & Sons, Ltd.

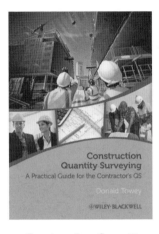

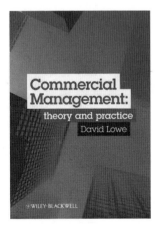

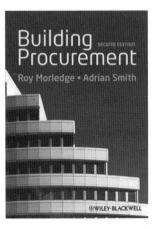

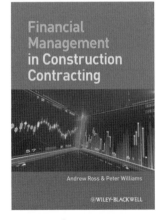